D0712962

WAVEFORMS

WAVEFORMS

A Modern Guide to Nonsinusoidal Waves and Nonlinear Processes

Homer B. Tilton

PRENTICE-HALL, INC.
Business & Professional Division
Englewood Cliffs, New Jersey

Prentice-Hall International, Inc., *London*
Prentice-Hall of Australia Pty. Ltd., *Sydney*
Prentice-Hall Canada Inc., *Toronto*
Prentice-Hall Hispanoamericana, S.A., *Mexico*
Prentice-Hall of India Private Ltd., *New Delhi*
Prentice-Hall of Japan, Inc., *Tokyo*
Prentice-Hall of Southeast Asia Pte. Ltd., *Singapore*
Whitehall Books Ltd., *Wellington, New Zealand*
Editora Prentice-Hall do Brasil Ltda., *Rio de Janeiro*

Library of Congress Cataloging-in-Publication Data

Tilton, Homer B.
Waveforms : a modern guide to nonsinusoidal waves and nonlinear processes.

Bibliography: p.
Includes index.
1. Analog electronic systems—Design and construction. 2. Electric waves. 3. Relaxation methods (Mathematics) I. Title.
TK7867.T495 1986 621.38′043 85-12427

ISBN 0-13-946096-9

Printed in the United States of America

This book is dedicated to the memory of my father, Benjamin Ellsworth Tilton.

ABOUT THE AUTHOR

Homer B. Tilton obtained his B.S. degree in Engineering Physics from Montana State University and holds graduate credit certificates from UCLA and the University of Arizona in engineering and physics. He served on the faculty of the Engineering Research Laboratory at the University of Arizona and has also been vice president of Optical Electronics, Inc.

Presently, Mr. Tilton is a member of the technical staff at Hughes Aircraft Company and proprietor of Visonics Laboratories, a research study effort concerned with visual electronics.

INTRODUCTION

This book is concerned with the mathematical representation and manipulation of nonsinusoidal waves such as squarewaves, pulse trains, sawtooth waves, and triangular waves. Balthasar van der Pol first referred to these kinds of periodic functions as *relaxation oscillations,* a name that is still used today. Consequently, we refer to functions having finite discontinuities and breaks as *relaxation functions,* and to the mathematics of nonsinusoidal waves and nonlinear processes as *relaxation analysis.*

There exists a smattering of relaxation analysis in the electronics literature, but there are many gaps; there appears to be no single source that pulls it all together. One source (Sheingold[1]) recognizes the relationship between clipping (or bounding) and center clipping (or dead zone). The term "discontinuous" is not always used according to its mathematically accepted definition. For example, we find $|x|$ referred to as a "discontinuous function," even though it is the derivative of $|x|$, not $|x|$ itself, which contains a discontinuity (at $x = 0$).

Also, there exists in the literature a proliferation of names for the same operations. Some of this we have indicated above. Clipping, bounding, limiting, and slicing all refer to the same operation.

In light of all this, what has been needed is a single comprehensive source for this mathematical discipline. This volume is an effort to fill

[1] D. H. Sheingold, ed., *Nonlinear Circuits Handbook,* 2nd ed. (Norwood, Mass.: Analog Devices, Inc., January 1976).

that need. Included herein is all such material within the author's purview (from existing literature) and new material that is required to join together the existing concepts and to extend them into other areas of immediate or potential usefulness.

This volume will show you the contemporary analog signal processing art from a mathematical standpoint. Always (except in the applications) the central object is the mathematical entity. Thus the expression sqr x *is* the sine-derived squarewave; and the vector $(y_1, y_2) = (\sin x, \cos x)$ *is* the mathematical spaceform called a circle.

But in addition to equations, this volume also uses block diagrams (and an occasional mechanical or electrical circuit) to illustrate principles, operations, and sequences of operations.

This book is organized into four parts. In the first part, Chapters 1 and 2 present basic concepts of relaxation analysis. Chapters 3, 4, and 5 in part two show how to write and manipulate relaxation functions, their derivatives, and their integrals. In the third part, Chapters 6, 7, and 8 deal with operations and processes dear to the heart of the analog circuit designer. Finally, in part four, Chapters 9, 10, and 11 deal with advanced concepts. A chapter-by-chapter synopsis follows.

Chapter 1 presents background, guidelines, basic rules, and definitions. It provides the foundation for the remainder of the book. If questions occur in the reader's mind concerning rules while reading the subsequent chapters, Chapter 1 should be briefly reviewed. The rules and definitions that are peculiar to this discipline are set down there for easy reference.

Step functions and delta functions (or impulse functions) are covered in Chapter 2. A familiarity with these mathematical objects is required to gain an understanding of the material in later chapters. The Dirac delta function assumes a central role in relaxation analysis, as it must, because it appears whenever a discontinuous function is differentiated. Useful properties of delta functions that are not required for an understanding of material in the initial chapters are developed in Chapter 9 and Appendix B.

Chapter 3 covers "compact" functional notation for squarewaves, triangular waves, sawtooth waves, and other nonsinusoidal waves. This chapter focuses on nonpiecewise representations, as these have been found to be more useful than piecewise ones.

Chapters 4 and 5 deal with differentiation and integration, respectively. Methods are presented that facilitate the differentiation and integration of virtually any relaxation function. These methods are illustrated by numerous examples and applications.

Wide pulses and pulse trains are considered in Chapter 6. Here it is shown how use of the Heaviside step function (introduced in Chapter

2) facilitates the representation of virtually any kind of rectangular pulse or pulse train. Triangular and sinusoidal pulses are also treated. Also shown in this chapter is how pulse functions can be used as gates.

In Chapter 7 it is shown how half-wave rectification can be represented in terms of the Heaviside step function. You will then see how variations of this expression can be used to perform clipping. A discussion of ramps, diode characteristics, and diode function generation follows naturally out of this.

Methods of representing and manipulating hysteretic functions are covered in Chapter 8. The various kinds of hysteresis are categorized. Again, the Heaviside step function plays a central role in this chapter in a specialized form as the "unit hystor." Hysteretic functions constitute a group of functions that gains in clarity by a detailed mathematical treatment.

Most of the functions and operations in this book have no more than finite discontinuities; however, Chapter 9 is devoted to functions with infinite discontinuities. This gives the volume a greater degree of mathematical completeness.

Chapter 10 discusses spaceforms—mathematical entities that use more than one single-valued function for their definition. Such spaceforms include Lissajous figures and rasters. A discussion of rasters leads naturally to a discussion of introductory principles of computer-generated displays. A joining of analog and digital concepts in the generation of raster-based computer-generated displays (rasterforms) is indicated. The material in this chapter is presented in such a way that extension beyond two- and three-dimensional spaceforms is facilitated. Examples are given of two-, three-, and four-dimensional spaceforms.

Even though the interpretation of functions of a complex variable as four-dimensional spaceforms dates back to Poincaré, this interpretation is rarely stressed. One purpose of Chapter 11 is to develop this idea; the material is designed to give the reader a new insight into this fascinating area.

In general, the methods of this book reveal the underlying interconnections among the many nonlinear processes of interest to the electronic design engineer. Many of these interconnections have been heretofore ignored because they were not apparent; but with their translation into the language of mathematics, the various interconnections are readily seen.

For example, the fact that we can integrate the pulse operation (puls) to obtain the clipping operation (clip) is obscure until one compares the defining expressions. As another example, the underlying reason that the frequency of $\sin^2 x$ is twice that of $\sin x$ is explained and extended to a general principle that permits many frequency-doubling expressions to be readily found. Finally, the appearance of staircase functions in the integrals of full-wave rectified sine and cosine waves is, perhaps, unexpected.

Besides serving as a powerful tool, relaxation analysis can provide the user with an insight that will not only assist in the design and analysis of nonlinear circuits and systems, but can also suggest new areas of design to be conquered.

In summary, the methods of relaxation analysis serve to complement and supplement those more traditional methods used by the designer: methods such as graphical analysis, frequency-domain analysis, and that time-honored method—trial and error.

But the analog circuit designer is not the only one affected by the methods of relaxation analysis. For example, the Dirac delta function is useful in the description of laser output because of its spectral narrowness. Therefore these methods are useful to the optical engineer.

In general, anyone engaged in or interested in information processing should find this book useful, for these techniques indicate methods of representing and manipulating information without regard to the medium. The medium need not be electrical or electronic. It can be mechanical, fluidic, optical, magnetic, even neural.

Homer Benjamin Tilton
Tucson, Arizona

HOW TO USE THIS BOOK

Before attempting to use the material on differentiation and integration (Chapters 4 and 5), the first three chapters should be thoroughly reviewed. Although much of the material in Chapters 6 through 11 can be used without mastery of differentiation and integration, these chapters also lean heavily on material in the first three chapters—especially Chapter 2. Generally speaking, applications information is given in the closing sections of the chapters.

This book is a reference, and the format is that of a *self-study manual;* such a format is advantageous when presenting a substantial amount of unique technical material, as is the case here. The main text presents and develops material in a logical sequence; at the same time, many of the relationships that are developed are collected together in the appendix for easy reference.

SPECIAL NOTE TO THE ANALOG CIRCUIT DESIGNER

An important step in the design process is the production of block diagrams. The experienced engineer relies to a large extent on intuition in producing these diagrams when system equations are not available. The methods presented in this book facilitate writing equations for nonlinear systems. The result is less reliance on intuition in producing block diagrams, because the relationship of a block diagram to the system equation is a direct one-to-one relationship.

Thus an important application of the methods presented in this book is in the production of block diagrams. Examples of this application are given throughout.

In this volume you will learn how some of the functions and operations that are useful in analog signal processing can be represented and manipulated mathematically. In the process, you will also see how some of these ideas can be used to extend and clarify some of the mathematics appearing in standard analysis.

While you will find a substantial variety of nonsinusoidal functions and nonlinear operations, there remains literally an infinite number of these still uncataloged. One purpose of this volume is to delineate some of the mathematical tools that are useful in the exploration of this territory.

ACKNOWLEDGMENTS

Special thanks are due to Professor William S. Bickel, whose continued support and encouragement were, in large part, responsible for this work being carried through to completion. I also wish to acknowledge the support and patience of my wife, Sue.

H. B. T.

TABLE OF CONTENTS

1

PROPERTIES OF DISCONTINUOUS FUNCTIONS

1.1 Prior Art: Heaviside's Operational Methods

Oliver Heaviside is generally credited with the introduction of the operational calculus, one form of which is the *Laplace transformation*. One thing that the development of the Laplace transformation did was to make it possible to work with nonsinusoidal periodic functions almost as if they were ordinary algebraic and transcendental functions.[1]

For example, the transform of the squarewave with period 2π is $(1/s)\tanh(\pi s/2)$; that of the triangular wave with period 2π is $(1/s^2)\tanh(\pi s/2)$; and that of the half-wave rectified sinewave with period 2π is $(s^2 + 1)^{-1}(1 - e^{-\pi s})^{-1}$. These expressions are all combinations of algebraic and transcendental functions in the variable s; but one must work in the transformed (s) domain to use them. The square, triangular, and rectified sine waves described above are graphed in Figs. 1.1, 1.2, and 1.3, respectively.[2]

In the Laplace transform, multiplication (division) by s corresponds to differentiation (integration) of the original function. This is illustrated by the relationship between the squarewave and triangular wave of Figs.

[1]Precise definitions of *sinusoidal* and *nonsinusoidal* are given in Chapter 3. For now, the dictionary definitions will suffice.

[2]The term "wave" is used synonymously with "periodic function." Thus *triangular wave* refers to a triangular periodic function.

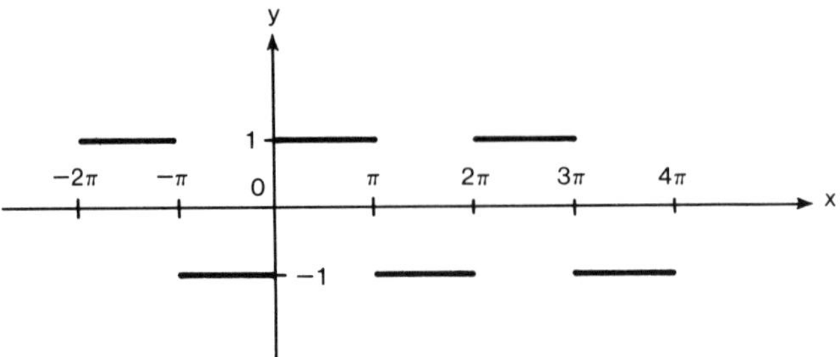

Figure 1.1 A squarewave.

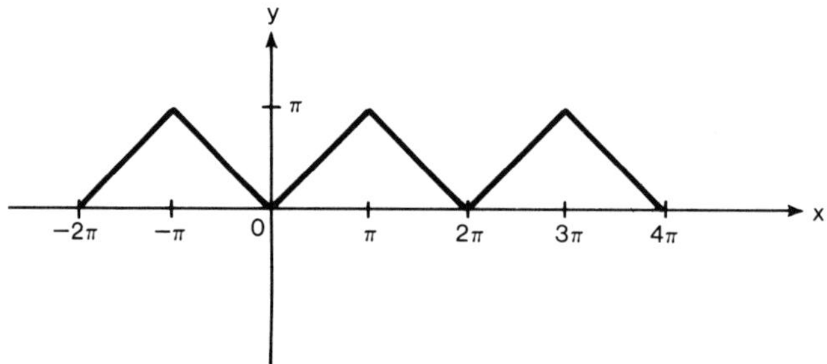

Figure 1.2 A triangular wave.

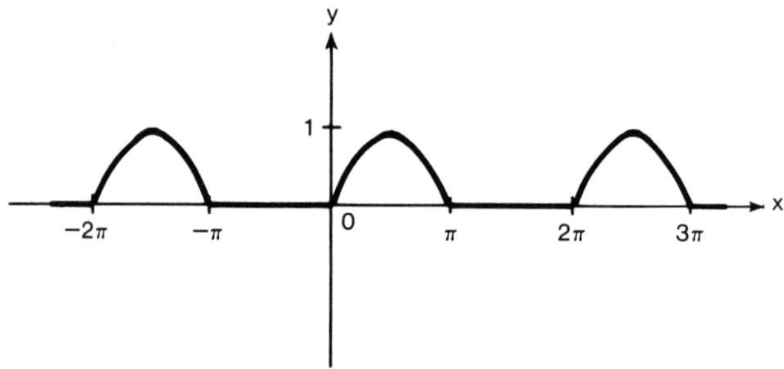

Figure 1.3 A half-wave rectified sinewave.

1.1 and 1.2, since that squarewave is the derivative of that triangular wave.

Three properties of the Laplace transform that contribute to its usefulness are as follows. First, integral transforms such as the Laplace transform have the property of transforming piecewise-continuous functions into continuous ones. (We will find later that this property is characteristic of integration in general.) Second, the Laplace transform has a unique inverse, except for an arbitrary *null function*. (Null functions are nonzero [or zero] but integrate like zero. We will encounter null functions throughout the present volume.) Third, if *time* is taken as the independent variable of the original function, then s can be interpreted as a complex frequency. Thus the Laplace transform permits the correspondence to be studied between *time domain* and *frequency domain* characteristics of linear systems. Unfortunately, these methods are not well-suited to nonlinear systems.[3]

Linear and nonlinear systems are contrastingly defined in terms of the superposition principle (it holds for linear systems only), in terms of the describing differential equations, or in terms of preservation of sinusoidal waveshapes (a linear system produces a sinusoidal output for a sinusoidal input). The third definition of nonlinearity best suits our immediate purposes: *A nonlinear system is one in which a sinusoidal input does not necessarily produce a sinusoidal output.* Thus linearity is included as a special case of nonlinearity. We do, indeed, find that the methods presented herein are applicable to linear systems and nonlinear systems equally.

Recently developed techniques (also traceable to Heaviside by way of the Heaviside step function and to Dirac by way of the Dirac delta function) make it possible to synthesize and manipulate squarewaves, triangular waves, and other broken and discontinuous functions directly by the use of nonoperational methods. Thus we are able to work with nonlinear systems; these methods therefore supplement the methods of the Laplace transformation. These methods are called by the name *relaxation analysis,* so called because they deal with squarewaves, sawtooth waves, etc., which are generated by relaxation oscillators.

The remaining sections of this chapter detail the fundamental considerations that form the foundations of relaxation analysis.

1.2 Absolute-Value and Principal-Value Operations

We begin this section with a brief review of notation and terminology.

If a quantity y (dependent variable) depends on another quantity x (independent variable) then we write symbolically

$$y = f(x) \tag{1.1}$$

where $f(\ )$ denotes an *operation* or *process* that acts on its *argument* (the quantity appearing within the parentheses, in this case x) to produce the *function* $f(x)$.

[3]An excellent discussion of these important aspects is given on pp. 1–3 of John G. Truxal, *Automatic Feedback Control System Synthesis* (New York: McGraw-Hill, 1955).

For example, if we let $f(\ ) \to \sin(\ )$, then Eq. (1.1) becomes

$$y = \sin x \tag{1.2}$$

or if we let $f(\ ) \to |(\ )|$, then Eq. (1.1) becomes

$$y = |x| \tag{1.3}$$

Further, if we let $u = u(x)$ be a u-function of x, then

$$y = g(u) \tag{1.4}$$

is a *function of a function*. For example, if $g(\ ) \to |(\ )|$ and $u(\ ) \to \sin(\ )$, then Eq. (1.4) becomes

$$y = |\sin x| \tag{1.5}$$

The above distinction between the concepts *operation* and *function* is maintained throughout the present volume. For example, we draw a distinction between the *step operation* and the *step function* discussed in Chapter 2. The latter is a particular function (or class of functions). The former acts on a function to produce a different function.

Equation (1.1) can be represented by the block diagram in Fig. 1.4. In the context of the block diagram, x is the *input*, y is the *output*, and $f(\ )$ is the *transfer characteristic* of the *function generator* (the five-sided block). We note that the direction of cause-effect is from left to right in the block diagram but from right to left in the equation it represents.

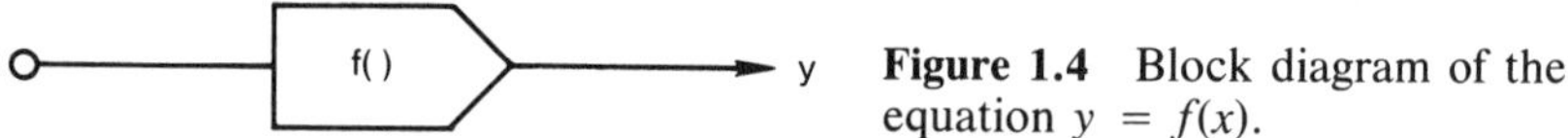

Figure 1.4 Block diagram of the equation $y = f(x)$.

Two elemental operations form the basis for relaxation analysis. They are the *absolute-value* and *principal-value* operations. Both of these are introduced in *standard analysis*[4], and it is assumed that the reader is familiar with them. Both act to discard or disregard certain information.

The absolute-value operation disregards sign of a number. The absolute-value operation, $|(\ )|$, acting on the argument x, produces the *absolute-value function,* $|x|$. This function is graphed in Fig. 1.5. By contrast, the principal-value operation is not so readily defined, as is shown in the following paragraphs. There is no *principal-value function* for a rectilinear variable (more correctly, the principal value of x is simply x), but there is a principal-value function for an angular variable. This case is discussed in Appendix D. Also, some writers refer to the principal values of inverse trigonometric functions as "principal-value functions."

The principal-value operation disregards all branches of a multiple-

[4]For the origin of this term see M. Davis and R. Hersh, "Nonstandard Analysis," *Scientific American,* 226, no. 6 (June 1972), 78–86.

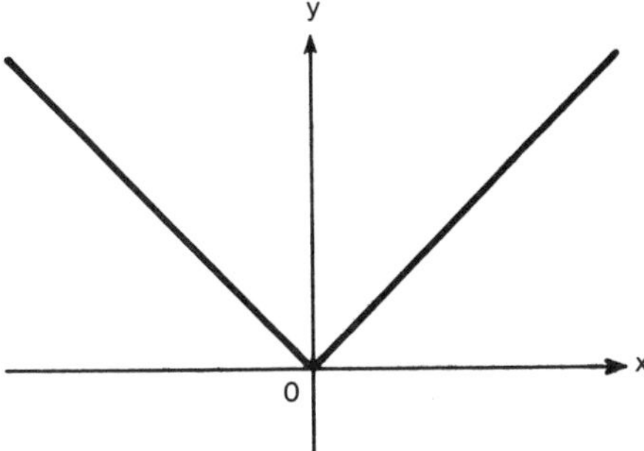

Figure 1.5 The absolute-value function.

valued function except the *principal branch,* thereby producing a single-valued function from the multiple-valued one. The principal-value operation, which we might denote by ($_{\equiv}$) in the case of inverse trigonometric functions (the triple underline indicating that the initial letter is to be capitalized) or by ($_+$) in the case of the square root, acts on a multiple-valued function to produce a single-valued function.

While radicals appearing in tables of integrals and in some textbooks are defined as the principal values of square roots, at other times $\sqrt{x^2}$ indicates the double-valued function graphed in Fig. 1.6. Korn and Korn might write the principal value of this function as $\underset{+}{\sqrt{x^2}}$.[5] We write it $(x^2)^{1/2}$.[6]

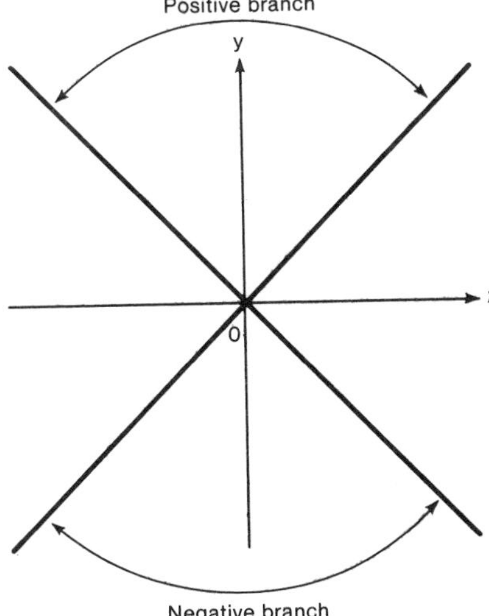

Figure 1.6 The function $\sqrt{x^2}$.

[5]See for example p. 9 of G. A. Korn and T. M. Korn, *Mathematical Handbook for Scientists and Engineers* (New York: McGraw-Hill, 1961).

[6]It is easy to see that $(x^2)^{1/2} \neq x$ because once x is squared, information of the sign of x is lost.

The principal value of $\sqrt{x^2}$ is defined as the principal branch of that function that in turn is defined as follows. We wish to have $(x^2)^{1/2} = [(-x)^2]^{1/2}$. There is only one way to define the two branches of $\sqrt{x^2}$ to accomplish this. This is as a *positive branch* (the V-shaped curve in Fig. 1.6) and a *negative branch* (the Λ-shaped curve). The principal branch is arbitrarily defined as the positive branch. Therefore $(x^2)^{1/2}$ is identical to $|x|$. In general we write $(\ )^{1/n}$ to mean the principal value of the nth root. The function x^a in general involves complex numbers. For this reason its general treatment is deferred to Chapter 11.

The principal branch of $\sqrt{x^2}$ is defined in an arbitrary manner. By contrast, the principal branches of the inverse trigonometric functions cannot be defined in an arbitrary manner because they are all determined by the definition of the prinicipal value of the natural logarithm. This is shown in Chapter 11. The resulting definitions of the principal branches for the six inverse trigonometric functions are listed in Table 1.1. Other important functions are listed in Table 1.1 also. In the table and elsewhere in this volume, boldfaced letters indicate complex variables. Otherwise, variables are real.

Also in the table and throughout the present volume, quantities that are angles, or that can be interpreted as angles, are specified in "natural"

TABLE 1.1 Principal Branches of Some Multiple-Valued Functions

MULTIPLE-VALUED FUNCTION	PRINCIPAL BRANCH	
	Symbol	Definition
• *Square root of square:*		
$\sqrt{x^2}$	$(x^2)^{\frac{1}{2}}$	$\|x\|$
• *Inverse trigonometric functions:*		
$\sin^{-1}x$	$\text{Sin}^{-1}x$	$-\frac{1}{2}\pi \leqslant \sin^{-1}x \leqslant \frac{1}{2}\pi$
$\cos^{-1}x$	$\text{Cos}^{-1}x$	$0 \leqslant \cos^{-1}x \leqslant \pi$
$\tan^{-1}x$	$\text{Tan}^{-1}x$	$-\frac{1}{2}\pi < \tan^{-1}x < \frac{1}{2}\pi$
$\text{ctn}^{-1}x$ ($\cot^{-1}x$)	$\text{Ctn}^{-1}x$	$0 < \text{ctn}^{-1}x < \pi$
$\sec^{-1}x$	$\text{Sec}^{-1}x$	$-\pi \leqslant \sec^{-1}x < -\frac{1}{2}\pi,\ 0 \leqslant \sec^{-1}x < \frac{1}{2}\pi$
$\csc^{-1}x$	$\text{Csc}^{-1}x$	$-\pi < \csc^{-1}x \leqslant -\frac{1}{2}\pi,\ 0 < \csc^{-1}x \leqslant \frac{1}{2}\pi$
• *Real power of a real variable:*		Definitions are given in Chapter 11
$\sqrt[1/a]{x}$	x^a	
• *Naperian logarithm of a real variable:*		
$\ln x$	$\text{Ln}\ x$	
• *Naperian logarithm of a complex variable:*		
$\ln \mathbf{x}$	$\text{Ln}\ \mathbf{x}$	
• *Angle of a complex variable:*		
$\arg \mathbf{x} = \theta$	$\text{Arg}\ \mathbf{x} = \Theta$	

units, i.e., *radians*. For example, $\mathrm{Cos}^{-1}(0) = \frac{1}{2}\pi$. The inverse trigonometric functions of Table 1.1 are graphed in Fig. 1.7. The heavy portions of the curves are the principal branches.

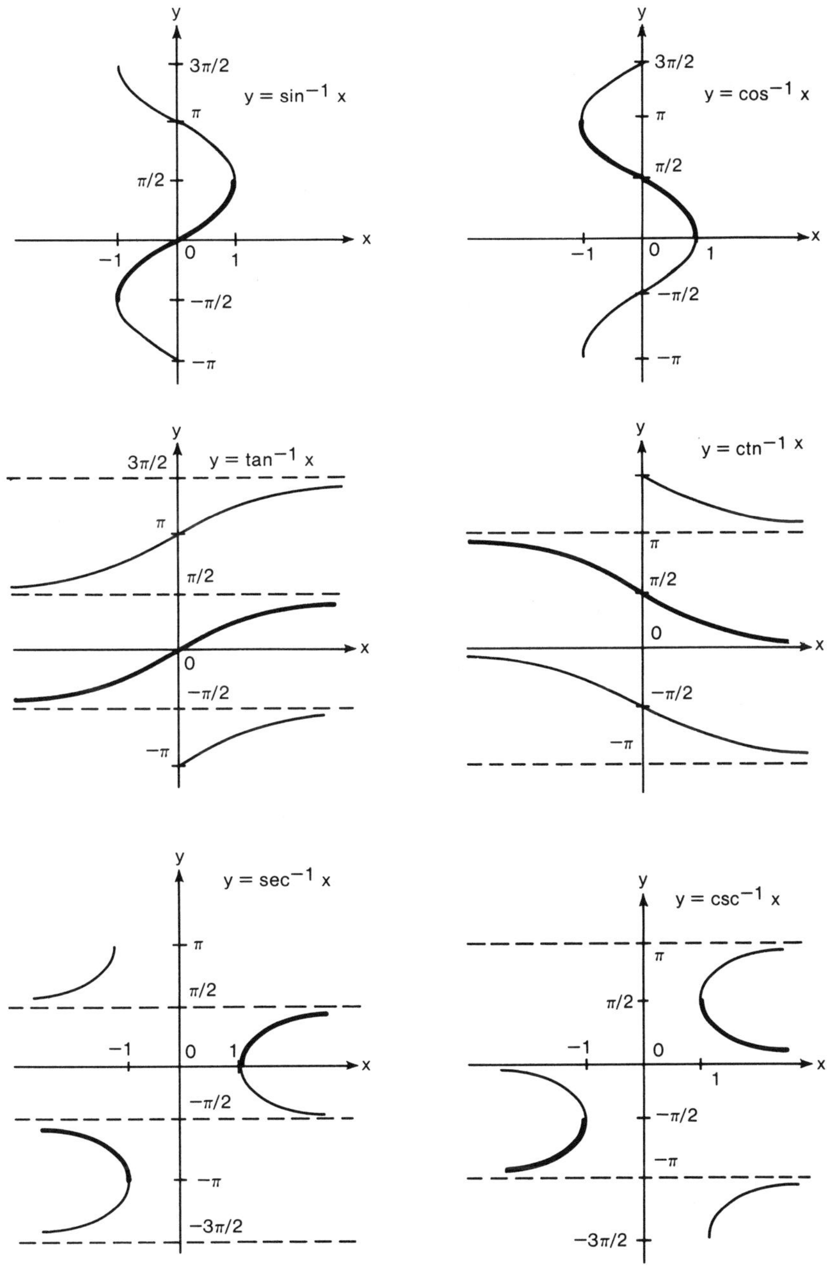

Figure 1.7 The inverse trigonometric functions.

Inverse trigonometric functions appearing in tables of derivatives and integrals are generally principal values.

Electronic circuits that generate inverse trigonometric functions produce single-valued functions, of necessity, although these need not be the principal values as defined above.

We group the special functions of relaxation analysis under the name *relaxation functions,* according to the following definition:

> *A relaxation function is any function formed from a function of standard analysis by applying any combination of absolute-value and principal-value operations.*[7]

The Dirac delta function is excluded from this class, although it plays an important role in relaxation analysis. As a general rule it will be found that relaxation functions are finitely discontinuous or have a finitely discontinuous nth derivative.

Van der Pol is generally credited with first using the term "relaxation oscillations" to refer to oscillations with discontinuous (or near-discontinuous) features.

1.3 Continuity as a Special Case of Discontinuity

The condition for continuity given in elementary calculus is as follows. A function $f(x)$ is said to be *continuous at* $x = a$ if

$$\lim_{x \to a} f(x) = f(a) \tag{1.6}$$

The limit of $f(x)$ must exist, and $f(a)$ must be at least definable. If Eq. (1.6) is not satisfied, $f(x)$ is said to be *discontinuous at* $x = a$. A more complete discussion of continuity can be found in virtually any text on the calculus.

Three kinds of discontinuities interest us in the present volume: finite discontinuities, infinite discontinuities, and cusps or breaks at infinity. The first of these we consider starting immediately. The second two we consider in Chapter 9.

A measure of discontinuity applicable to finitely-discontinuous functions is the *jump*, r, with

$$r = \lim_{x \to a+} f(x) - \lim_{x \to a-} f(x) \tag{1.7}$$

[7] Actually only the principal-value operation need be used, since the absolute value of a function can be written as the principal value of the square root of its square, as shown in the preceding paragraphs.

It will be recalled that $x \to a+$ $(x \to a-)$ is read "$x \to a$ from more-positive ($-$negative) values." Such limits are said to be *unilateral*. If $f(x)$ is continuous at $x = a$, then $r = 0$. In this sense the property of continuity is a special case of the property of discontinuity.

According to Eqs. (1.6) and (1.7), $f(x) = x$ and $f(x) = |x|$ are both continuous at $x = 0$ (and everywhere else). But there is an obvious difference between the two functions. Namely, the derivative of x is continuous throughout but the derivative of $|x|$ is discontinuous at $x = 0$. To distinguish between the two kinds of continuity, type x functions (with continuous first derivative) will be called *continuous-smooth,* and type $|x|$ functions (with discontinuous first derivative) will be called *continuous-broken*. These names are sometimes shortened to simply *smooth* and *broken*. We say that $|x|$ has a *break* or *breakpoint* at $x = 0$.

When we say a function is "continuous," we mean it is continuous throughout. When we say a function is "broken," we mean it is continuous throughout and has one or more isolated breaks. (It is important to remember that a broken function is also a continuous function.) When we say a function is "finitely discontinuous," we mean it is continuous throughout except for one or more isolated finite jumps. Such a function is also called *piecewise continuous*. Functions are generally single valued unless otherwise stated.

1.4 How to Represent Discontinuous Functions Simply

Relaxation analysis is largely an algebra of discontinuous functions. In nonoperational standard analysis, a discontinuous function is customarily represented as a limiting case of a continuous function, as a piecewise collection of continuous functional segments, or as the sum of an infinite number of continuous functions. By contrast, in relaxation analysis a discontinuous function is represented directly as a legitimate function in its own right. This is done in such a way as to establish a connection to standard analysis by way of the absolute-value and principal-value operations.

The *step function* can be considered to be the elementary discontinuous function since its discontinuity is its most apparent property. A representation of a step function that appears in nonoperational standard analysis is $\tanh(nx)$, with n assumed to have an indefinitely large value. One property of this step function is that

$$|x|\tanh(nx) = x \tag{1.8}$$

Solving Eq. (1.8) for $\tanh(nx)$ gives

$$\tanh(nx) = x/|x| \tag{1.9}$$

except possibly at $x = 0$. Indeed, it is this latter form, $x/|x|$, that we find most useful as a representation of this particular step function.

Another way to represent this step function is in *piecewise form* as

$$y = -1, \qquad x < 0 \tag{1.10a}$$

$$y = 1, \qquad x > 0 \tag{1.10b}$$

Piecewise representations are generally characterized by the requirement to use *conditional inequalities* ($x < 0$ and $x > 0$ in the above example) as an adjunct to the piecewise functional expressions. Primarily for this reason their usefulness is limited.

The *sawtooth wave* is another discontinuous function, one that is periodic. Such a function is normally represented in nonoperational standard analysis as the sum of an infinite series of sine functions—a *Fourier series*. By contrast, in relaxation analysis sawtooth waves are represented by expressions of the type $\mathrm{Tan}^{-1}\tan x$, where Tan^{-1} is the principal value of the multiple-valued inverse tangent function, $\tan^{-1}$. The reader can easily verify that

$$\left.\begin{aligned} \mathrm{Tan}^{-1}\tan x &= x, \qquad -\tfrac{1}{2}\pi < x < \tfrac{1}{2}\pi \\ &\neq x \text{ elsewhere}^{8} \end{aligned}\right\} \tag{1.11}$$

It is not difficult to show that this function is the sawtooth wave graphed in Fig. 1.8.

In Chapters 2 and 3 it is shown how the forms $|u|$, $u/|u|$, and $\mathrm{Tan}^{-1}\tan u$ can be extended and combined to synthesize useful representations for a wide variety of discontinuous and broken functions.

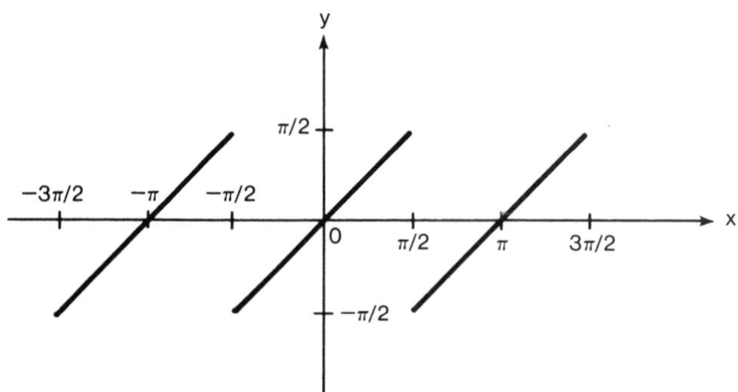

Figure 1.8 A sawtooth wave.

[8]The points at $-\frac{1}{2}\pi$, $\frac{1}{2}\pi$, and other points of discontinuity are excluded from consideration in Eq. (1.11). The special consideration those points require is given in Chapter 3.

1.5 Why Not Use Piecewise Representations?

The present approach focuses on nonpiecewise representations of functions. Some reasons for this follow. Basically, it is found that the abandonment of piecewise notation permits and facilitates the exploration of territory that otherwise is difficult (if not impossible) to explore analytically.

When frequencies and phases of nonsinusoidal waves involved in a calculation are related in a simple way, piecewise representations many times suffice. At other times a *compact* (nonpiecewise) representation is more convenient to write and to work with. For example, it is almost always preferable to write the full-wave-rectified sinewave in its compact form, $y = |\sin x|$, instead of in the piecewise form

$$\begin{aligned} y &= \sin x, \qquad 2n\pi < x \leqslant (2n+1)\pi \\ y &= -\sin x, \qquad (2n+1)\pi < x \leqslant (2n+2)\pi \\ n &= \ldots,\ -1,\ 0,\ 1,\ 2,\ 3,\ \ldots \end{aligned} \tag{1.12}$$

Simple operations involving periodic functions can frequently be handled in a piecewise representation. However, not all operations lend themselves well to piecewise treatment. One example is the class of operations that produces a frequency change, e.g., frequency-doubling operations. Another example is the class of operations that produces a nonperiodic function from a periodic one, e.g., integration of functions whose average value is not zero. Frequency-doubling operations are treated in Chapter 3. Integrations that produce nonperiodic functions from periodic ones are encountered in Chapter 5 and elsewhere.

By using a compact notation that is connected to standard analysis, we are able to draw from a large body of knowledge that has accumulated over the centuries. Thus, for example, we are able to integrate the sawtooth wave in the form $\text{Tan}^{-1}\tan x$ by using the standard integration formula

$$\int (u^2+1)^{-1}\text{Tan}^{-1}u\, du = \tfrac{1}{2}(\text{Tan}^{-1}u)^2 \tag{1.13}$$

This is shown in Chapter 5.

For these reasons and others, the focus of the present volume is on compact representations based on absolute-value and principal-value operations.

1.6 Brushing the Boundaries of That No Man's Land Known as "Division by Zero"

Relaxation analysis, of necessity, brushes the boundaries of that no man's land known as "division by zero" in two areas. First, we encounter so-called *indeterminate expressions* of the form 0/0. This area has been extensively explored in standard analysis.

The following is an excerpt from Widder[9]:

The determination of the limit

$$\lim_{x \to c} \frac{f(x)}{g(x)} \quad (1)$$

where $f(c) = g(c) = 0$ is traditionally referred to as the evaluation of the indeterminate form 0/0. This phraseology is misleading in as much as division by zero is undefined.[10] But the evaluation of the limit (1) is fundamental to the calculus. For example, the problem arises in the very definition of the derivative of a function

$$f'(x_0) = \lim_{\Delta x \to 0} \frac{f(x_0 + \Delta x) - f(x_0)}{\Delta x}$$

for both numerator and denominator tend to zero with Δx. In computing the derivative of a given elementary function (and in evaluating other expressions of the form 0/0), some algebraic reduction or other device must always be employed to avoid the indeterminate character.

Second, division by an infinitesimal is implicit in the definition of the Dirac delta function. This area has been explored in mathematical physics where the usefulness of the Dirac delta function is well-established.

1.7 Unilateral Limits and Other Important Concepts

When working with discontinuous functions it is important to specify direction of progression along the curve (cf. the notations $x \to a+$ and $x \to a-$ used earlier). This is a consideration that is unimportant with continuous functions but crucial to the proper treatment of discontinuous ones.

For example, the "value" of $\tan x$ "at" $x = \frac{1}{2}\pi$ (either $+\infty$ or $-\infty$) depends on the direction of progression. *It certainly is not zero there.* We generally ignore that point in standard analysis, simply saying the limits $\lim_{x \to \pi/2+} \tan x$ and $\lim_{x \to \pi/2-} \tan x$ "do not exist." Even though the two limits "do not exist," *each limit is the negative of the other* in the sense

[9] D. V. Widder, *Advanced Calculus* (Englewood Cliffs, N.J.: Prentice-Hall, Inc., 1947), p. 216.

[10] It is also misleading in that such expressions are frequently determinate.

that $\lim_{x\to\pi/2} \dfrac{\tan x}{\tan(\pi - x)} = -1$. Use is made of this kind of relationship in Chapter 9 where the "complete" derivative of tan x is found. This derivative applies everywhere, including at $x = \frac{1}{2}\pi$ and all other points of discontinuity.

As another example, the value of $x/|x|$ at $x = 0$ is not obvious; we might be tempted to say it is zero. However, this would not be realistic in view of the fact that it is customary—and justifiably so—to assign the value $+1$ at $x = 0$ to $(\sin x)/x$ and x/x. Nor can we afford the luxury of simply ignoring the point at $x = 0$.

It is shown in Chapters 2 and 3 how the concept of unilateral limits permits resolution of the question concerning the value to be assigned to functions at a point of discontinuity.

In addition to "continuous" and "discontinuous," functions can be classified as "even," "odd," or other. A function $y = f(x)$ is *even* if $f(-x) = f(x)$. Examples of even functions are $y = |x|$, $y = 1/|x|$, and $y = \cos x$. Even functions are *symmetrical* about (or with respect to) the y-axis.

A function $y = g(x)$ is *odd* if $g(-x) = -g(x)$. If $g(x)$ is continuous at $x = 0$, then we must have $g(0) = 0$. Examples are $y = x$ and $y = \sin x$. If $g(x)$ is discontinuous at $x = 0$, then $g(0)$ need not be zero. Examples are $y = 1/x$ and $y = \text{ctn}\, x$. Odd functions are *antisymmetrical* about the y-axis. If $g(0) \neq 0$, the antisymmetry is said to be "global" for reasons given in Chapter 2.

If a function $y = h(x)$ is neither even nor odd, then it is asymmetrical about the y-axis.

Finally, definitions and terminology relating to periodic functions are set down for reference.

A function $p(x)$ is *periodic in x with period a* if $p(x + a) = p(x)$ for all x-values.[11] We note that a function that is periodic with period a is also periodic with period $2a$, $3a$, etc. We usually take the smallest possible value as "the period." This is the *fundamental period*. That portion of a periodic function within an interval equal to the fundamental period is called a *cycle*. The reciprocal of the fundamental period is the *fundamental frequency*, or simply the *frequency*. For sinusoids, 2π times the frequency is called *angular frequency*. Simplification of analysis sometimes results if we take the period as a multiple of the fundamental period. Cf. Example 4.4.

[11] The case $p(x)$ = constant is generally excluded although a constant may be said to be a *degenerate* periodic function.

If a function $q(x)$ is periodic in x with fundamental period a and is such that $q(x + \frac{1}{2}a) = -q(x)$, then it possesses *half-wave symmetry;* it is said to be *antiperiodic* in addition to being periodic.

In the text, certain relationships are flagged by a quiver (→→→). These relationships are (1) implications or findings of theorems, (2) conclusions of proofs, or (3) definitions. The reader should not expect, however, that all "important" relationships are so flagged.

With the foundation established, we now begin our journey into the fascinating world of relaxation analysis with a detailed discussion of step functions, staircase functions, and delta functions.

2

THE ANATOMY OF STEP, STAIRCASE, AND DELTA FUNCTIONS

2.1 The Signum Function of x Is the Sign of x

Step functions are functions of the type

$$\text{step } u = E \qquad \text{when } u < 0 \tag{2.1a}$$

$$\text{step } u = P \qquad \text{when } u > 0 \tag{2.1b}$$

where $u = x - a$, and E, P, and a are constants with $P \neq E$.[1] They are discontinuous functions with a discontinuity called a *step* at $u = 0$ ($x = a$) whose jump or *rise* is $P - E$. The value of a step function at the step is variously defined in the literature. Often it is left undefined. A step is an illustration of a finite discontinuity.

The *signum function* is -1 for negative values of its argument and $+1$ for positive values. We now define a step function, *sgn*, which has these properties.

$$\rightarrow\rightarrow\rightarrow \quad \operatorname{sgn} x \equiv x/|x| \tag{2.2}$$

This function is graphed in Fig. 2.1. Just as $|x|$ extracts the absolute value or *magnitude* of x, sgn x extracts its sign or polarity. It is clear that sgn x is $+1$ when $x > 0$ and -1 when $x < 0$. Therefore it is a step function with step of rise 2 at $x = 0$.

The value of sgn x at $x = 0$ is to be established. We recognize that

[1]The symbols E and P are used in anticipation of the introduction of a pictorial notation for the general step function given in the next section.

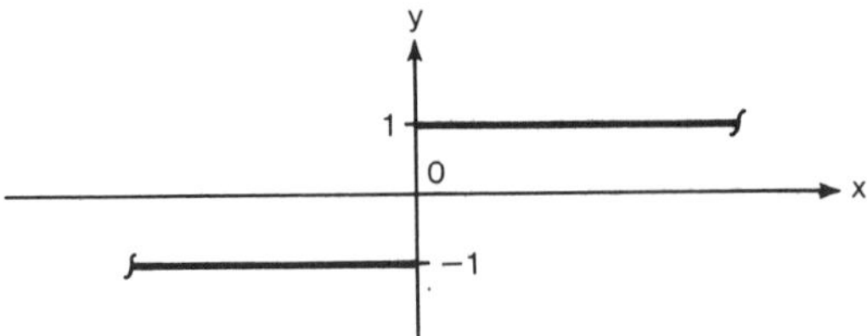

Figure 2.1 The signum function.

$\operatorname{sgn} x = x/|x|$ at $x = 0$ is an indeterminate expression of the form 0/0, and proceed to evaluate it by the method of limits. We write

$$\operatorname{sgn}(0) = \lim_{x \to 0} x/|x| \tag{2.3}$$

Approaching $x = 0$ first from negative values of x, i.e., for *positive progression,* the limit is

$$\lim_{x \to 0-} x/|x| = \lim_{x \to 0-} x/(-x) = -1 \tag{2.4}$$

Similarly, approaching $x = 0$ from positive values of x, i.e., for *negative progression,* the limit is

$$\lim_{x \to 0+} x/|x| = \lim_{x \to 0+} x/x = 1 \tag{2.5}$$

That is, *the value of sgn x at the step depends on the direction of progression along the curve*. This is an illustration of *infinitesimal hysteresis* wherein the value of a function at a single point depends on the direction of progression.[2] The dots in Fig. 2.2 indicate the position of the point at $x = 0$. The arrows indicate the direction of progression.

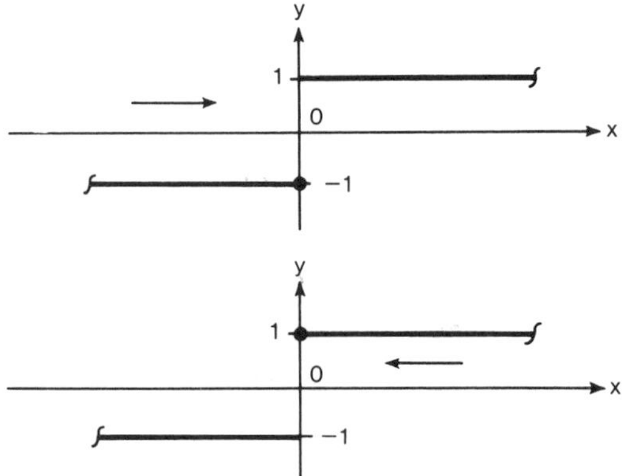

Figure 2.2 Signum function: The point at $x = 0$.

[2]In general, when $r \neq 0$ in Eq. (1.7), no unique value exists for $f(a)$. We say that $\lim_{x \to a-} f(x)$ gives *the value of $f(a)$ for progression in the $+x$ direction,* and $\lim_{x \to a+} f(x)$ gives *the value of $f(a)$ for progression in the $-x$ direction.* In this sense an *infinitesimal hysteresis* exists at $x = a$ wherein the value of $f(a)$ depends on the direction of progression.

Because of the existence of infinitesimal hysteresis at $x = 0$, the signum function is said to have *global* antisymmetry. In general an odd (even) function with infinitesimal hysteresis is said to have "global" antisymmetry (symmetry) because such a function is antisymmetrical (symmetrical) in the usual sense only when both directions of progression are considered.

Since 1 is its own reciprocal (as is -1), it follows that sgn x is its own reciprocal; it can also be written for real x-values as[3]

$$\operatorname{sgn} x = |x|/x \tag{2.6}$$

Also,

$$\frac{\operatorname{sgn} x}{\operatorname{sgn} x} = \operatorname{sgn} x \operatorname{sgn} x = |\operatorname{sgn} x| = 1 \tag{2.7}$$

While sgn x is a function, sgn( ) is an operation. Therefore we refer both to the signum function (meaning the step function) and the signum operation (meaning an operation that extracts the sign of its argument).

It is noted that the sgn x function as defined in Eq. (2.2) is the same as Korn and Korn's[4] Sgn x function except at $x = 0$.

The signum operation can also be combined with other operations to produce operations that do not simply extract sign. An example is the *spring preload* operation considered by Savant.[5] We represent it by *prld* and define

$$\operatorname{prld}_a u = u + a \operatorname{sgn} u \tag{2.8}$$

A graph of $\operatorname{prld}_a u$ is shown in Fig. 2.3.

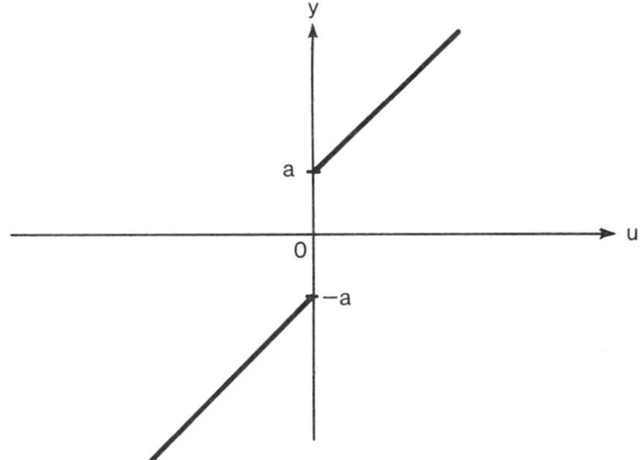

Figure 2.3 The preload operation.

[3]The situation is different for complex x-values as is shown in Chapter 11.

[4]G. A. Korn and T. M. Korn, *Mathematical Handbook for Scientists and Engineers* (New York: McGraw-Hill, 1961), p. 741.

[5]See problem 10-1 on p. 348 of C. J. Savant, Jr., *Basic Feedback Control System Design* (New York: McGraw-Hill, 1958).

In electronic parlance the step operation is called *amplitude comparison*. Other electronic operations closely related to the step operation are *zero-crossing detection* and *infinite clipping*. These are discussed in Sections 6.6 and 7.5.

Electronic devices that perform the step operation are called *two-level comparators, amplitude comparators,* or simply *comparators*. Practical comparators are designed with finite hysteresis (as opposed to infinitesimal hysteresis) to provide immunity from incidental noise. Finite hysteresis results from the use of regenerative feedback. Such comparators are also called *Schmitt triggers* for the inventor of the vacuum-tube version.

2.2 The Heaviside Step Function

The *Heaviside step function* is zero for negative values of its argument and +1 for positive values. We now define a step function, S, which has these properties. It is

$$\rightarrow\rightarrow\rightarrow \quad S(x) \equiv \tfrac{1}{2} + \tfrac{1}{2}\operatorname{sgn} x \tag{2.9}$$

Its value at $x = 0$ is either 0 or +1, depending on the direction of progression along the curve. The function $S(x)$ is graphed in Fig. 2.4. It exhibits infinitesimal hysteresis because sgn does.[6]

Some properties of the Heaviside step function are

$$S(x)S(x) = |S(x)| = S(x) \tag{2.10a}$$

$$S(x) + S(-x) = 1 \tag{2.10b}$$

$$S(x) - S(-x) = \operatorname{sgn} x \tag{2.10c}$$

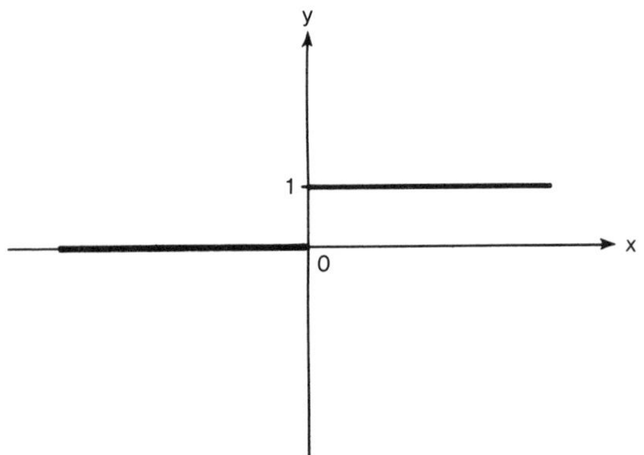

Figure 2.4 The Heaviside step function.

[6]Even though other definitions than Eq. (2.9) are given for the Heaviside step function, we will nevertheless refer to S as defined in Eq. (2.9) as the Heaviside step function.

We now introduce a pictorial notation for step functions. It is $\mathrm{st}\underline{E}|\overline{P}$ (read "step from E to P") where

$$\mathrm{st}\underline{E}|\overline{P}\,(u) = \mathrm{step}\,u \tag{2.11}$$

with step u being defined as in Eqs. (2.1a) and (2.1b), step (0) $= E$ for positive progression, and $= P$ for negative progression.

Using this notation the Heaviside step function and the signum function are written

$$S(x) = \mathrm{st}\,\underline{0}|\overline{1}\,(x) \tag{2.12a}$$

$$\mathrm{sgn}\,x = \mathrm{st}\,\underline{-1}|\overline{1}\,(x) \tag{2.12b}$$

Also, we can write step u in terms of S(u) and sgn u as

$$\mathrm{step}\,u = E + (P - E)S(u) \tag{2.13a}$$

$$\mathrm{step}\,u = \tfrac{1}{2}(P + E) + \tfrac{1}{2}(P - E)\,\mathrm{sgn}\,\mathrm{u} \tag{2.13b}$$

Example. Write the step function $\mathrm{st}\,\underline{0.1}|\overline{10}\,(x - 1000)$ (a) in terms of the Heaviside step function; (b) in terms of the signum function; (c) in terms of the absolute-value function.

Answer. $\mathrm{st}\,\underline{0.1}|\overline{10}\,(x - 1000) =$
(a) $0.1 + 9.9\,\mathrm{S}(x - 1000)$;
(b) $5.05 + 4.95\,\mathrm{sgn}(x - 1000)$;
(c) $5.05 + 4.95\,\dfrac{x - 1000}{|x - 1000|}$

It is clear that the notation $\mathrm{st}\,\underline{\mathrm{E}}|\overline{\mathrm{P}}\,(x - a)$ is adequate to fully define any desired step function. The following elementary relationships are presented without proof.

The product of a step function and a constant is the step function

$$K\,\mathrm{step}\,u = \mathrm{st}\,\underline{KE}|\overline{KP}\,(u) \tag{2.14}$$

The sum of a step function and a constant is the step function

$$K + \mathrm{step}\,u = \mathrm{st}\,\underline{K + E}|\overline{K + P}\,(u) \tag{2.15}$$

The reciprocal of a step function, if one exists, is the step function

$$(\mathrm{step}\,u)^{-1} = \mathrm{st}\,\underline{E^{-1}}|\overline{P^{-1}}\,(u) \tag{2.16}$$

The sum of two step functions is a step function if both steps are at the same point. Where step′u means $\mathrm{st}\,\underline{E'}|\overline{P'}\,(u)$ it is

$$\mathrm{step}\,u + \mathrm{step}'\,u = \mathrm{st}\,\underline{E + E'}|\overline{P + P'}\,(u) \tag{2.17}$$

The product of two step functions is a step function if both steps are at the same point. It is

$$\mathrm{step}\,u\,\mathrm{step}'u = \mathrm{st}\,\underline{EE'}|\overline{PP'}\,(u) \tag{2.18}$$

The ratio of two step functions, if it exists, is a step function if both steps are at the same point. It is

$$\frac{\text{step } u}{\text{step}' u} = \text{st } \underline{E/E'}|\overline{P/P'}\,(u) \tag{2.19}$$

The reflection in the y-axis of $y = \text{step}(x - a)$ is the step function

$$\text{refl}_y[\text{step}(x - a)] = \text{step}(a - x) = \text{st } \underline{P}|\overline{E}\,(x + a) \tag{2.20}$$

Finally, st $\underline{A}|\overline{A}\ (u) = \text{A}$

Examples

1. Find the reciprocal of $1 + 2\text{sgn } u$

$$(1 + 2\text{sgn } u)^{-1} = [\text{ st } \underline{-1}|\overline{3}\,(u)]^{-1}$$

$$= \text{st } \underline{-1}|\overline{1/3}\,(u) = -1 + 4/3S(u)$$

$$= -1/3 + 2/3 \text{ sgn } u$$

Checking: $(1 + 2\text{sgn } u)(-1/3 + 2/3 \text{ sgn } u) = 1$

2. Find the reciprocal of $1 + \alpha S(u)$

$$[1 + \alpha S(u)]^{-1} = \text{st } \underline{1}|\overline{1/(\alpha + 1)}\,(u)$$

$$= 1 - \frac{\alpha}{\alpha + 1} S(u)$$

Checking: $[1 + \alpha S(u)]\left[1 - \frac{\alpha}{\alpha + 1} S(u)\right] = 1$

3. Find the product of $S(u)$ and step u

$$S(u) \text{ step } u = st\ \underline{0}|\overline{1}\,(u)\ st\ \underline{E}|\overline{P}(u)$$

$$= st\ \underline{0}|\overline{P}\,(u) = \text{P } S(u)$$

2.3 Staircase Functions Contain a Series of Steps

Staircase functions are functions consisting of multiple steps. The portion of the function between adjacent steps, and the size of that portion, is called the *tread.* A staircase function consisting of an infinite number of steps of equal rise and equal tread is called a *regular staircase function.* The particular regular staircase function with symbol $\sigma_a(x)$ is shown in Fig. 2.5. Its rise and tread both equal a. It exhibits infinitesimal hysteresis. It can be written

$$\sigma_a(x) = a\sigma_1(x/a) \tag{2.21}$$

where $\sigma_1(x/a)$ can be found from the doubly-infinite series

$$\begin{aligned}\sigma_1(u) &= S(u - \tfrac{1}{2}) + S(u - 3/2) + S(u - 5/2) + \cdots \\ &\quad - S(-u - \tfrac{1}{2}) - S(-u - 3/2) - S(-u - 5/2) - \cdots\end{aligned} \tag{2.22}$$

A better form for σ is presented in Chapter 3.

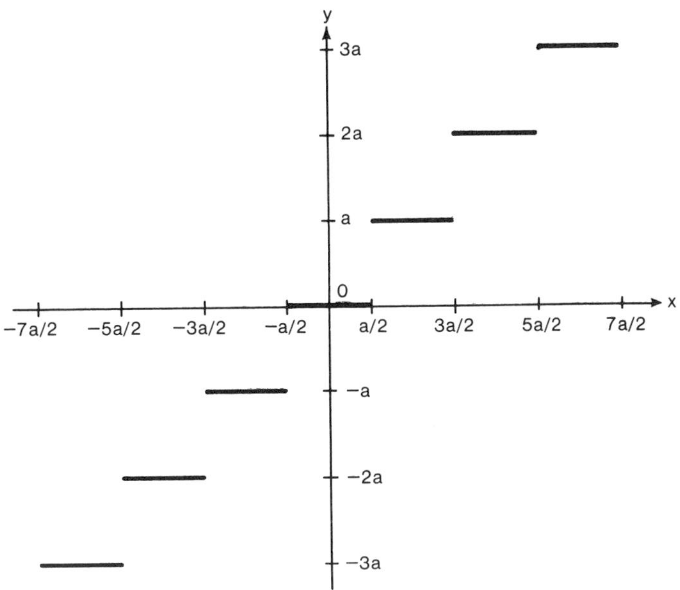

Figure 2.5 A regular staircase function.

An *alternating staircase function* is one in which adjacent rises are of opposite signs. An example is diagrammed in Fig. 2.6. The absolute value of this particular alternating staircase function is the same as the absolute value of $\sigma_a(x)$. Expressions for this function and for $\sigma_a(x)$ are developed in Chapter 3.

Staircase generators exist as specialized devices for generating repetitive staircase functions of time. In Chapter 3 we show how to write expressions for such functions.

A particular staircase function of interest is the *two-step staircase function* shown in Fig. 2.7. Its equation is

$$y = S(x - a) + S(x - b) \tag{2.23}$$

The function of Eq. (2.23) can be generalized by substituting a general step function for the two S-functions, as follows:

$$y = \text{step}\,(x - a) + \text{step}\,(x - b) \tag{2.24}$$

The electronic device that performs this operation is a *three-level comparator*. It is also called a *dead-zone comparator*.

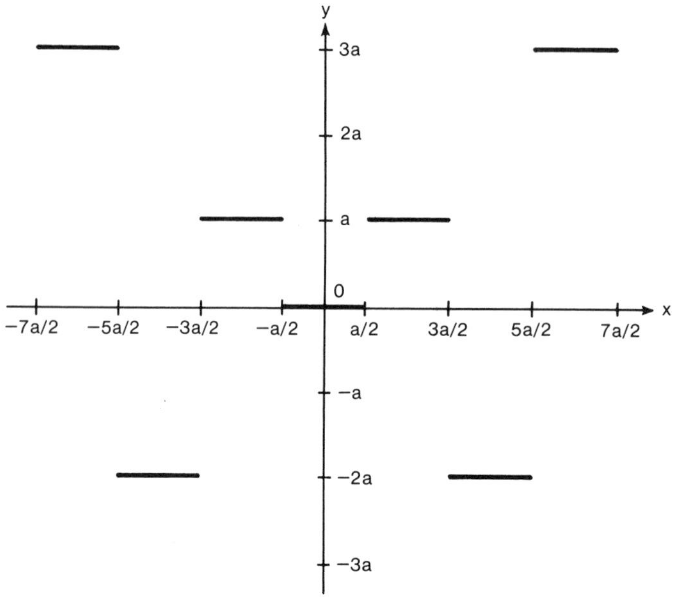

Figure 2.6 An alternating staircase function.

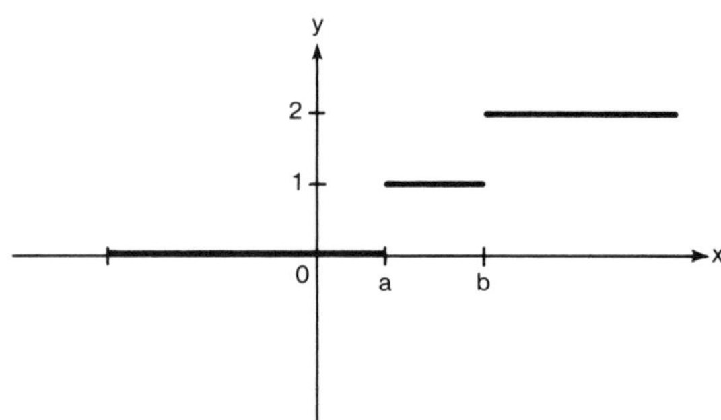

Figure 2.7 A two-step staircase function.

Another staircase function that is of interest is the alternating two-step staircase called a unit *window function, filter function,* or *pulse function.* We assign it the symbol *puls* and write it

$$\text{puls}_a^b x \equiv S(x - a) - S(x - b) \tag{2.25}$$

This function is characterized by a unit step up at $x = a$ and a unit step down at $x = b$. If $a < b$, it is a positive unit pulse. If $a > b$, it is a negative unit pulse. These two cases are diagrammed in Fig. 2.8.

This function is important enough to have a major portion of a separate chapter (Chapter 6) devoted to it.

The electronic device that performs the window operation is called a *window comparator*.

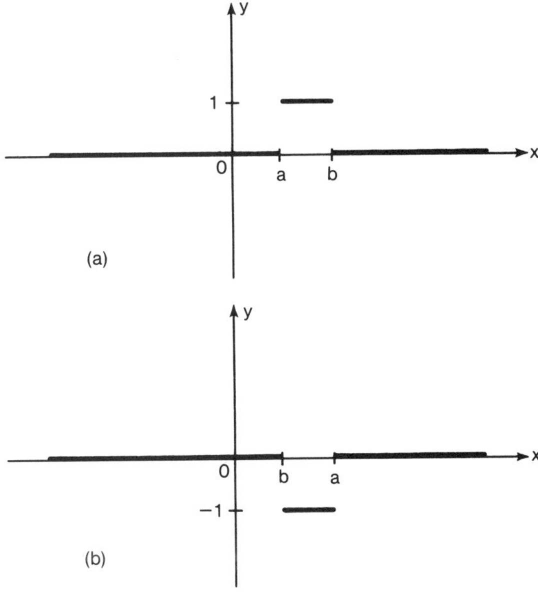

Figure 2.8 Pulse functions.

Another kind of staircase function is the *stile function*. The *second-order stile function* shown in Fig. 2.9 has the equation

$$y = S(x + 3/2) + S(x + \tfrac{1}{2}) - S(x - \tfrac{1}{2}) - S(x - 3/2) \qquad (2.26)$$

The window function is a *first-order stile function*. Higher order stile functions can be found by extending the pattern suggested by Eqs. (2.25) and (2.26).

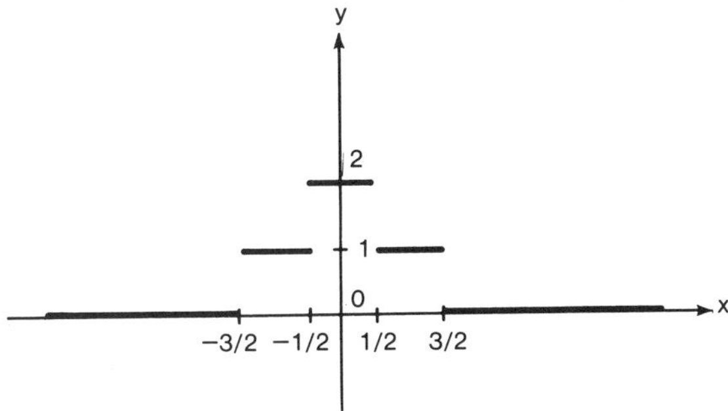

Figure 2.9 A second-order stile function.

In general, any staircase function can be formed as a linear combination of step functions.

Still another kind of staircase function is the *integer function*. We discuss two kinds of integer functions. The first kind, which we represent by square brackets, [ ], is a regular staircase function defined as

$$y = [x] = \sigma_1(x - \tfrac{1}{2}) \tag{2.27}$$

It is graphed in Fig. 2.10. This integer function is asymmetrical with respect to the y-axis. It exhibits infinitesimal hysteresis because σ does.

The second kind of integer function, which we represent by *Int*, is the same as the first kind for positive values of the argument and for negative progression; but otherwise it differs. It is graphed in Fig. 2.11.

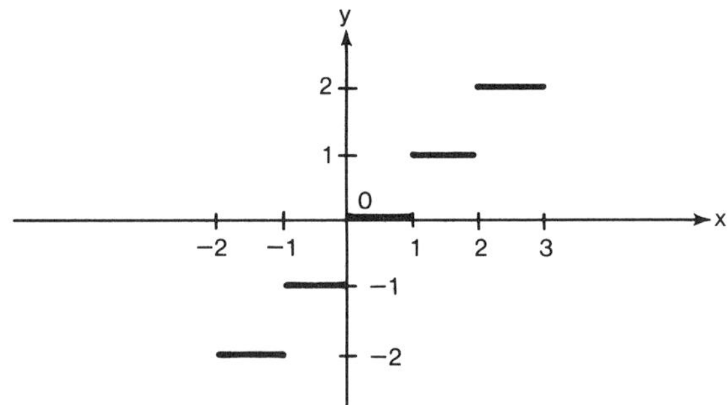

Figure 2.10 An integer function.

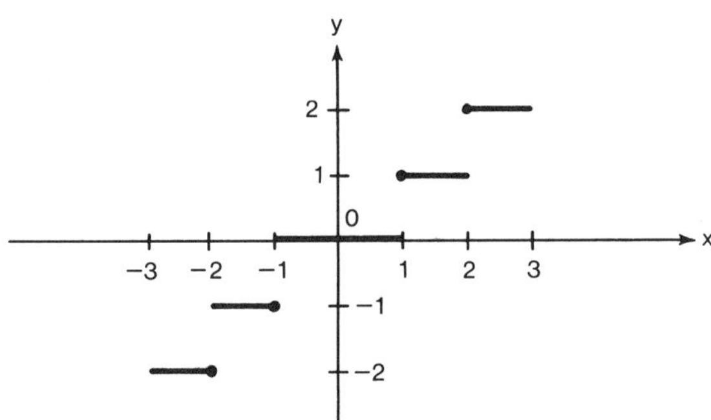

Figure 2.11 Another integer function.

The function Int(x) does not exhibit infinitesimal hysteresis. It is not expressible in terms of σ.

A function that is related to the Int function is the *fractional function,* or *frac*. It is equal to

$$\text{frac}(x) = x - \text{Int}(x) \tag{2.28}$$

The functions Int(x) and frac(x) are not relaxation functions. They commonly appear on scientific pocket calculators, and are included here solely in the interest of completeness.

2.4 The Dirac Delta Function: Is It Really a Function?

Delta functions are everywhere zero except for one or more pulses of infinitesimal width—*delta pulses*. The particular delta function with symbol δ is the *Dirac delta function*. Delta functions are also called *impulse functions* and *singularity functions*.

These mathematical objects are traditionally called functions; but are they really functions? From Arfken[7] with permission of the author and the publisher:

> From these defining equations $\delta(x)$ must be an . . . infinitely thin spike. . . . the problem is that *no such function exists*[8] in the usual sense of function. . . .
>
> A way out of this difficulty is provided by the theory of distributions. . . . $\delta(x)$ is labeled a distribution (not a function).

These mathematical objects are therefore properly called *delta pulse distributions;* nevertheless we will call them delta functions as is the fashion.

The indefinite integrals of delta functions are functions even though the delta functions themselves are not. Therein lies their usefulness.

The Dirac delta function is traditionally defined in terms of the properties that $\delta(x) = 0$ when $x \neq 0$ and

$$\int_{-a}^{b} \delta(x)dx = 1, \qquad a, b > 0 \tag{2.29}$$

However, those two properties follow from the more fundamental relationship

$$\rightarrow\rightarrow\rightarrow \quad dS(x) = \delta(x)dx \tag{2.30}$$

Integrating Eq. (2.30) we obtain

$$\int \delta(x)dx = S(x) + C \tag{2.31}$$

from which Eq. (2.29) immediately follows. Also it is apparent from Eq. (2.30) that $\delta(x)$ is everywhere zero except at $x = 0$, where it must be infinite, i.e., *$\delta(x)$ is a delta function with delta pulse of infinite amplitude at $x = 0$.*

[7] G. Arfken, *Mathematical Methods for Physicists*, 2nd ed. (New York: Academic Press, 1970), pp. 413–415.

[8] The emphasis is Arfken's.

A formal definition of the Dirac delta function is sometimes given as

$$\delta(x) = \frac{1}{2\pi}\int_{-\infty}^{\infty} e^{ixy}dy \tag{2.32}$$

where e is the base of natural logarithms and $i^2 = -1$.

Another way to view the Dirac delta function is as the limit of a finite pulse whose area remains constant at unity in the limit. An example of such a *delta sequence function* is given in Appendix B. Cf. the final section of Appendix B.

2.5 Other Delta Functions; Delta Pulse Trains

Other delta functions can be defined in terms of the Dirac delta function. For example, $2\delta(x)$ is a delta function which, when integrated, produces a step function with rise 2 at $x = 0$. We can write

$$\int 2\delta(x)dx = 2S(x) + C = \text{sgn } x + C' \tag{2.33}$$

A quantity called the *content* of a delta function is defined as the rise of the step that is produced upon its integration. For example, the rise of the step function produced by integrating $\delta(x)$ is 1 at $x = 0$; therefore, the content of $\delta(x)$ is said to be 1 at $x = 0$. The content of $2\delta(x)$ is 2 at $x = 0$, and so on.

It is convenient to graph delta functions in terms of content rather than value. Thus, the Dirac delta function is represented as in Fig. 2.12.

It is apparent that

$$\delta(\text{sgn } x) = 0 \tag{2.34}$$

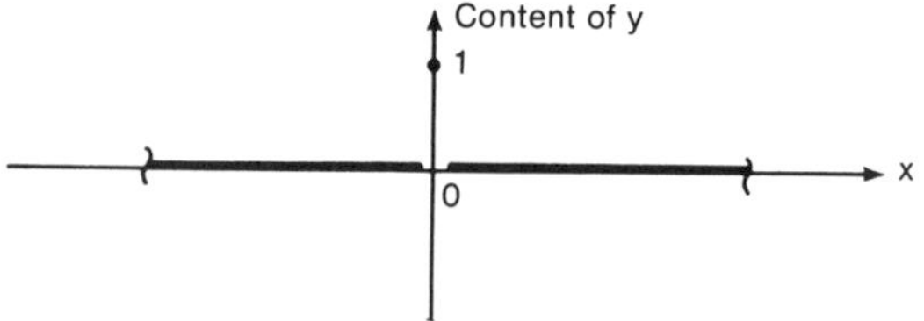

Figure 2.12 The Dirac delta function represented in terms of its content.

because sgn x never takes on the value zero. Other properties of delta functions are now given.

It is easy to show from Eq. (2.29) that $\delta(-x) = \delta(x)$ and in general, with a being a constant or variable that is independent of x,

$$\delta(a - x) = \delta(x - a) \tag{2.35}$$

If $f(x)$ is single valued at $x = 0$, then $f(x)\delta(x) = f(0)\delta(x)$. (Obviously true when $x = 0$. Also when $x \neq 0$, $f(x)\delta(x) = 0$, and $f(0)\delta(x) = 0$.) It follows that[9]

$$\int f(x)\delta(x)dx = f(0)S(x) + C \tag{2.36}$$

It is now easy to show that

$$\int_{-\infty}^{\infty} f(x)\delta(x)dx = f(0) \tag{2.37}$$

$$\int g(x)\delta(x - a)dx = g(a)S(x - a) + C \tag{2.38}$$

$$\int_{-\infty}^{\infty} g(x)\delta(x - a)dx = g(a) \tag{2.39}$$

where g(x) must also be single valued at $x = a$.

Inspection of Eq. (2.36) shows that $f(x)\delta(x)$ is a delta function with content $f(0)$ at $x = 0$. Similarly, inspection of Eq. (2.38) shows that $g(x)\delta(x - a)$ is a delta function with content $g(a)$ at $x = a$.

It follows from Eq. (2.36) that

$$\int x\delta(x)dx = C \tag{2.40}$$

The delta function $x\delta(x)$ integrates like zero although it is not literally zero. We call this kind of delta function a *vestigial delta function*. It belongs to the general class of *null functions*. This kind of delta function is examined later in this volume.

Other kinds of delta functions that we will find useful are *delta pulse trains*. A delta pulse train is a series of delta pulses that occur with regular frequency. An example is

$$y = \delta(\sin x) \tag{2.41}$$

This delta pulse train is represented graphically in Fig. 2.13. A pulse occurs every time sin x crosses zero. This is a *sine-derived delta pulse*

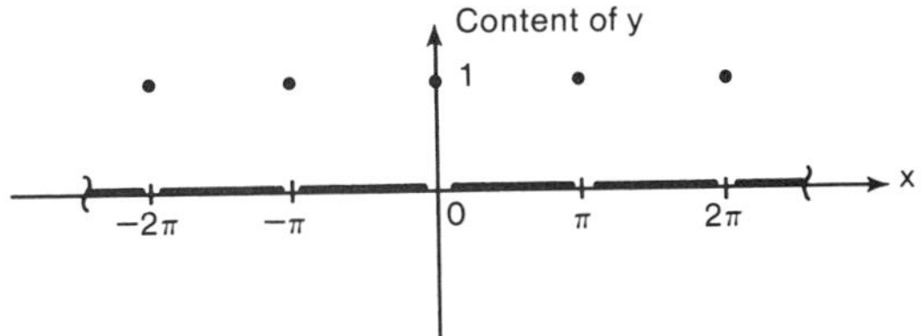

Figure 2.13 A sine-derived delta pulse train.

[9]Stieltjes considered such integrals in the form $\int f(x)dS(x)$ before the invention of the Dirac delta function.

train. We note that the fundamental frequency of $\delta(\sin x)$ is twice that of $\sin x$.

The integral of a delta pulse train is a staircase function. For example,

$$\pi\int \delta(\cos x)dx = \sigma_\pi(x) + C \tag{2.42}$$

because each delta pulse produces a step when integrated. (We might call $\sigma_\pi(x)/\pi = \int\delta(\cos x)dx$ a *cosine-derived regular staircase function.*) Proof is deferred to Chapter 4. Compare Eq. (2.42) with Eq. (2.31).

The expression

$$y = \cos x\, \delta\, (\sin x) \tag{2.43}$$

is a *sine-derived alternating delta pulse train*. It is diagrammed in Fig. 2.14. When integrated, the pulse train of Eq. (2.43) produces the function $S(\sin x)$, that is,

$$\int \cos x\, \delta(\sin x)dx = \mathrm{S}(\sin x) + C \tag{2.44}$$

because each positive (negative) delta pulse produces a step of rise $+1$ (-1). This is a special kind of alternating staircase function called a *square pulse train*. We consider such functions further in Chapter 6. The square pulse train $\mathrm{S}(\sin x)$ is graphed in Fig. 2.15. It is easy to prove Eq. (2.44) by substituting u for $\sin x$.

From Eq. (2.9) it is seen that $S(\sin x)$ can also be written $\frac{1}{2} + \frac{1}{2}$ $\mathrm{sgn}(\sin x)$; where $\mathrm{sgn}(\sin x)$ is the sine-derived squarewave discussed in Chapter 3. Therefore we can also write Eq. (2.44) as $2\int\cos x\, \delta(\sin x)dx = \mathrm{sgn}(\sin x) + C$.

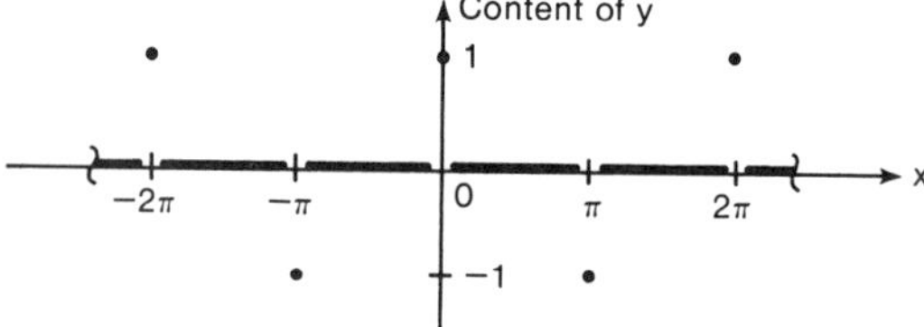

Figure 2.14 An alternating delta pulse train.

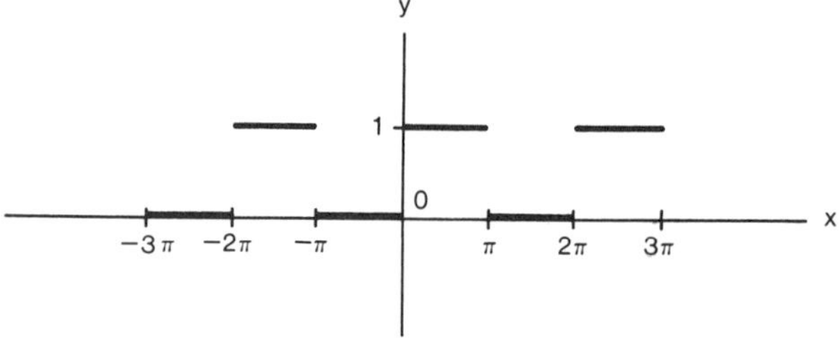

Figure 2.15 A square pulse train.

So far we have considered arguments $u = u(x)$ for δ that are continuous at $u = 0$ and that cross the x-axis with unit slope, i.e., $|du/dx| = 1$ when $u = 0$. Other possibilities are: (1) functions that are discontinuous at $u = 0$, (2) functions that are tangent to the x-axis so that they do not cross the x-axis (such as $u = x^2$), and (3) functions that cross at other than unit slope. The first possibility is not considered in this book except when such discontinuity means $u \neq 0$. Cf. Eq. (2.34). The second two possibilities are encompassed by the material in the following paragraphs.

As an example, we let $u = 2x$, which has a zero-crossing slope of two. Intuitively, the "width" of this delta pulse would seem to be half that of $\delta(x)$, so that its content relative to x would be one-half. The content is found from

$$\int \delta(2x)dx = \tfrac{1}{2}\int \delta(u)du = \tfrac{1}{2}S(u) + C = \tfrac{1}{2}S(x) + C \tag{2.45}$$

to be one-half as expected. This property can be generalized as

$$\delta(u) = |du/dx|^{-1}\delta(x - a) \tag{2.46}$$

where a is such that $u(a) = 0$.

The reader is cautioned that Eq. (2.46) is true only in the sense that the *content* of $\delta(u)$ is equal to the *content* of $|du/dx|^{-1}\delta(x - a)$. Differences remain between those two delta pulses that are examined in Appendix B.

As a result of the foregoing, $\delta(2 \sin \tfrac{1}{2}x)$ is a delta pulse train whose pulses have unit content, whereas the pulses of $\delta(\sin \tfrac{1}{2}x)$ have contents of two units. Also, the pulses of $\delta(\sin x + a)$ have unit content only when $a = 0$.

Finally, we introduce the *Kronecker delta* and show how it is related to the Dirac delta function. The Kronecker delta is a mathematical object one form of which is δ_{mn} where m and n are indices representing integers. It is defined as

$$\delta_{mn} = 1, \qquad m = n \tag{2.47a}$$

$$\delta_{mn} = 0, \qquad m \neq n \tag{2.47b}$$

It is shown in Appendix B that the delta function $|u|\delta(u) = \delta^0(u)$ also has these properties when $u = m - n$ or $u = n - m$. That is,

$$\delta^0(m - n) = \delta_{mn}; \qquad m, n \text{ integers} \tag{2.48}$$

However, $\delta^0(u)$ is defined for a continuum of arguments so that in general

$$\delta^0(u) = 1, \qquad u = 0 \tag{2.49a}$$

$$\delta^0(u) = 0, \qquad u \neq 0 \tag{2.49b}$$

Therefore δ^0 can be considered to be a generalization of the Kronecker delta that applies for any pair of real numbers m and n, not just for integers.

Also, δ^0 is a unit amplitude *infinitely-narrow window function*. It is a *vestigial delta function* and a *null function*. Other properties of δ^0 are listed in Appendix C.

It is also a *coincidence detector* because it is unity when m and n coincide (are equal) and zero otherwise. An electronic coincidence detector is equivalent to a narrow window comparator. Coincidence detection finds use in raster-based computer generated displays and elsewhere. Some applications of δ^0 are discussed in Chapters 7, 10, and 11.

If m and n are allowed to take on only the two values 0 and 1, then δ_{mn} performs the same operation as an *exclusive-nor gate* (XNOR-gate) in the system of *binary digital logic*.

2.6 Applications

One application of the delta function is in the description of monochromatic light. Continuous wave (cw) laser light approximates an ideal delta function of radiation. For example, the output, B, of a photoreceptor that is illuminated by monochromatic laser light is found as follows.

Let the photoreceptor output be given by

$$B = F\left[\int_0^\infty H_\lambda(\lambda)\phi(\lambda)d\lambda\right] \tag{2.50}$$

where λ is wavelength, $\phi(\lambda)$ is the relative spectral response function of the photoreceptor, $H_\lambda(\lambda)$ is the spectral irradiance—i.e., the spectral density of radiant power ("flux") $H(\lambda)$ falling on the photoreceptor so that $H_\lambda(\lambda) = dH(\lambda)/d\lambda$, and F is the transfer characteristic of the photoreceptor. The block diagram of Eq. (2.50) is shown in Fig. 2.16.

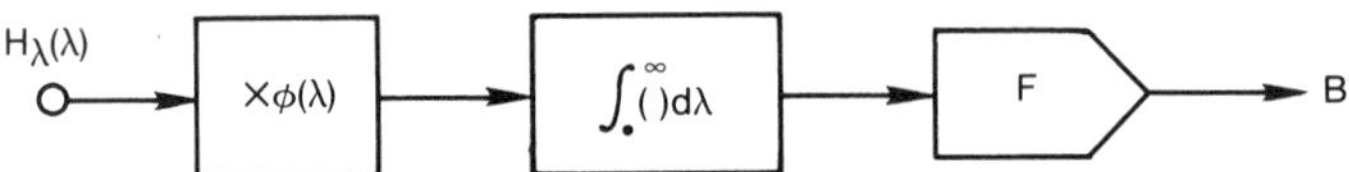

Figure 2.16 Block diagram of the photoreceptor equation.

Let the laser output at the photoreceptor be characterized by irradiance H and wavelength λ_0. Its spectral irradiance function at the photoreceptor is closely approximated by the delta function

$$H_\lambda(\lambda) = H\delta(\lambda - \lambda_0) \tag{2.51}$$

Finally,

$$B = F\left[H\int_0^\infty \phi(\lambda)\delta(\lambda - \lambda_0)d\lambda\right] = F[H\phi(\lambda_0)] \tag{2.52}$$

gives the photoreceptor output when its input is the laser radiation of Eq. (2.51).

Delta functions are also useful for describing the following quantities:

1. Space charge density function of an ideal point, line, or plane—approximated by an isolated electron in the case of a point charge
2. An ideal impulsive force—approximated when two steel balls collide
3. The voltage pulse in an ideal deflection yoke due to a sawtooth of current—approximated by the "flyback pulse" in a television flyback transformer
4. Frictional forces in an ideal mechanism whose static friction differs from its kinetic friction—approximated by real mechanisms.

The first of these can be found in virtually any modern text on electromagnetic theory. The second and third are examined in Section 4.4 of Chapter 4. The fourth is examined in Subsection 7.6.2 of Chapter 7.

Of the five applications above, four deal with delta functions that have finite contents. The remaining one (number 4) deals with a vestigial delta function. At the other end of the scale, "super delta functions," whose contents are infinite, find application in the differentiation of infinitely-discontinuous functions. This case is treated in Chapter 9.

Also, in Appendix B an extended discussion is given of powers and derivatives of the Dirac delta function.

Applications for step operations occur in pulse generation using comparators and in performing full- and half-wave rectification without the use of diodes. These applications are described in later chapters, notably Chapters 6 and 7.

2.7 Comment

The step and staircase functions of this chapter are elementary relaxation functions. Other elementary relaxation functions are presented in the following chapter.

3

THE SYNTHESIS OF NONSINUSOIDAL WAVES

3.1 Full-Wave Rectified Sinewave

If $y = A \sin(Bx + \phi)$ where A, B, and ϕ are constants, then y is a *sinusoidal function of x* and we say it is a *sinusoid*. The sinewave (sin x) and cosinewave ($\cos x$) are examples of sinusoids. If $y = p(x)$ is periodic and continuous and has the general form of a sinusoid, we say it is a *sinuous wave*. Thus, $\exp(\sin x)$ is a sinuous wave but it is not a sinusoid.

If a periodic function is not of the form $A \sin(Bx + \phi)$ and cannot be put in that form, it is *nonsinusoidal*. Examples of nonsinusoidal waves are $\exp(\sin x)$, $\tan x$, $\operatorname{ctn} x$, $\sec x$, and $\csc x$. The last four are of limited use in electronics because they are not everywhere finite.

A nonsinusoidal periodic function frequently encountered in electronics is the *full-wave rectified sinewave*. It is a relaxation function because it can be formed by applying the absolute-value operation to a transcendental function. It is considered next.

Consider the absolute value of u:

$$|u| = (u^2)^{1/2} \tag{3.1}$$

traditionally, or from Chapter 2,

$$|u| = u \operatorname{sgn} u \tag{3.2}$$

The function $y = |x|$ is graphed in Chapter 1 (cf. Fig. 1.5). It is a nonperiodic relaxation function. It is an even or symmetric continuous-broken function.

The absolute value is also called the *magnitude*[1] or *numerical value*.

If $u = \sin x$, Eq. (3.1) becomes

$$|\sin x| = (\sin^2 x)^{1/2} \tag{3.3}$$

This is a periodic relaxation function, and a nonsinusoidal function, of which four cycles are shown in Fig. 3.1. These four cycles are made from but two cycles of sin x. It is a continuous-broken function with an infinite number of breaks. It is a *full-wave rectified sinewave*. The absolute-value operation in electronic terminology is called *full-wave rectification*.

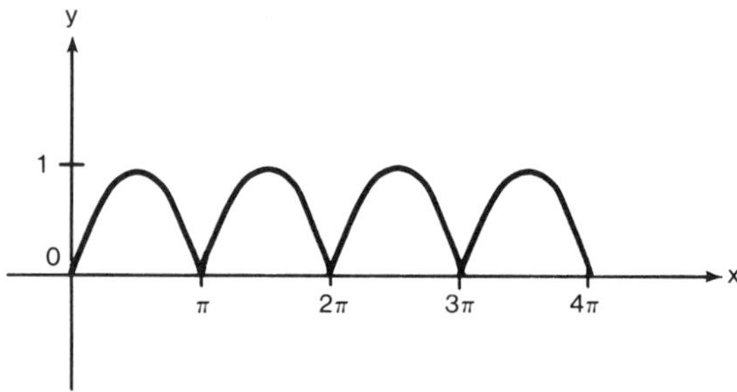

Figure 3.1 Full-wave rectified sinewave.

3.2 Squarewaves

From Eq. (3.2) we can write

$$|\sin x| = \sin x \operatorname{sgn}(\sin x) \tag{3.4}$$

where sgn(sin x) is the *squarewave* diagrammed in Fig. 3.2. We assign this squarewave the symbol *sqr* and write

$$\rightarrow\rightarrow\rightarrow \quad \operatorname{sqr} x \equiv \operatorname{sgn}(\sin x) \tag{3.5}$$

It is a function with an infinite number of finite discontinuities. It is a special kind of alternating staircase function.

It is seen that

$$\begin{aligned} \operatorname{sqr} x &= 1 && \text{when } \sin x > 0 \\ \operatorname{sqr} x &= -1 && \text{when } \sin x < 0 \end{aligned} \tag{3.6}$$

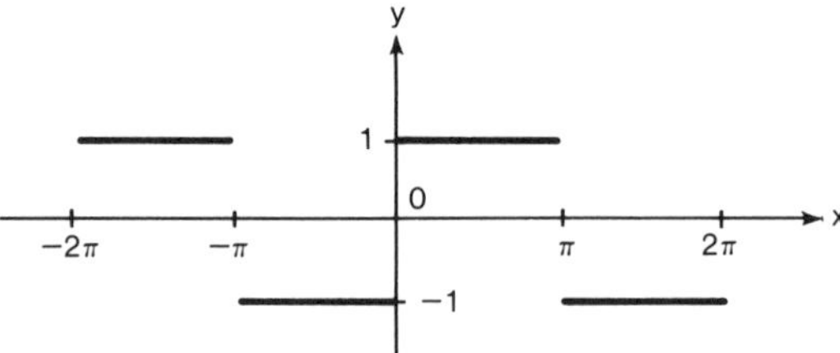

Figure 3.2 Sine-derived squarewave.

[1] A convenient way to read $|u|$ is *mag(u)* for *magnitude of u*.

When $\sin x = 0$, $\text{sqr}\, x$ is either $+1$ or -1 depending on the branch of the curve and the direction of progression along the curve, as shown in Fig. 3.3. The points at the discontinuities are represented by dots in Fig. 3.3. The direction of progression is indicated by the arrows. This is another illustration of infinitesimal hysteresis. The function $\text{sqr}\, x$ is an odd function with global antisymmetry.

It is clear that the full-wave rectified sinewave can be restored to a normal sinewave by multiplying it by $\text{sqr}\, x$. That is,

$$\sin x = \text{sqr}\, x\, |\sin x| \tag{3.7}$$

Other relationships are

$$\frac{\text{sqr}\, x}{\text{sqr}\, x} = \text{sqr}\, x\, \text{sqr}\, x = |\text{sqr}\, x| = 1 \tag{3.8}$$

$$\text{sqr}(-x) = -\text{sqr}\, x \tag{3.9}$$

The function $\text{sqr}\, x$ is a *sine-derived squarewave*. A *cosine-derived squarewave* can also be defined as

$$y = \text{sgn}(\cos x) \tag{3.10}$$

We can now write an expression for the alternating staircase function of Fig. 2.6. It is

$$y = -\text{sgn}\left(\frac{\pi}{a} \cos \frac{ax}{\pi}\right) \sigma_a(x)\, \text{sgn}\, x \tag{3.11}$$

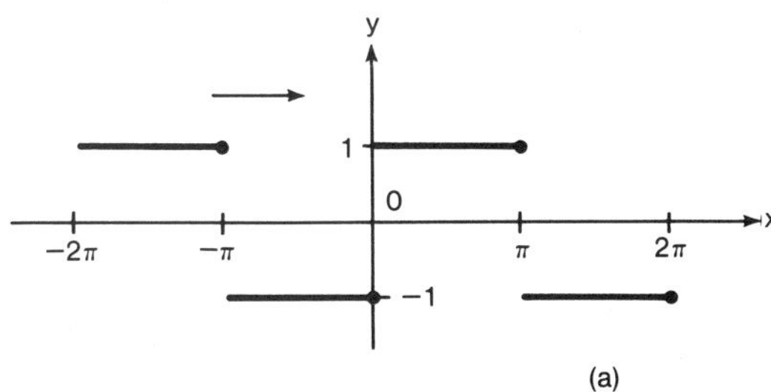

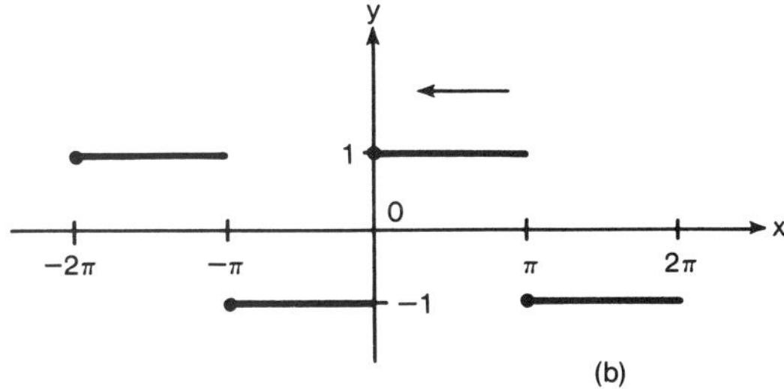

Figure 3.3 Squarewave: The points at the discontinuities.

3.3 Triangular and Sawtooth Waves

We now introduce the *triangular wave, tri,* represented graphically in Fig. 3.4. It has the property that

$$\text{tri}\, x = |x| \qquad \text{when } -\pi < x \leqslant \pi \tag{3.12}$$

and repeats periodically with period 2π. It is defined as

$$\rightarrow\rightarrow\rightarrow \quad \text{tri}\, x \equiv \text{Cos}^{-1}\cos x \tag{3.13}$$

where Cos^{-1} indicates the principal value of $\cos^{-1}$. It is a *cosine-derived triangular wave.*

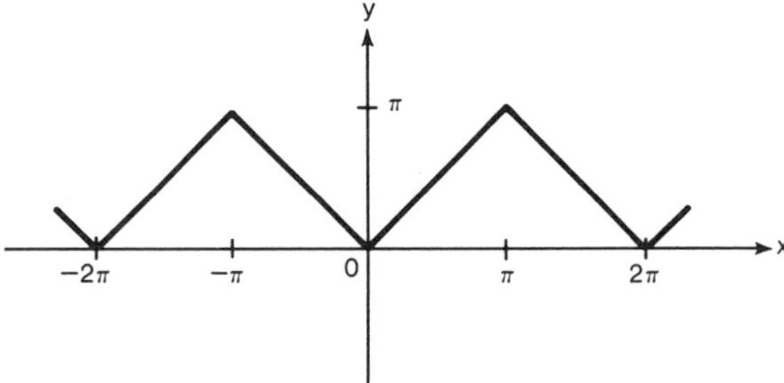

Figure 3.4 Cosine-derived triangular wave.

The reader should take a moment to verify that Eq. (3.13) is, indeed, the function diagrammed in Fig. 3.4, by recalling that the principal value of $\cos^{-1}u$ lies on the principal branch, which in turn is defined as that branch lying between zero and pi. That is, between $x = 0$ and $x = \pi$, $\text{Cos}^{-1}\cos x = x$; elsewhere $\text{Cos}^{-1}\cos x \neq x$. The function tri x is an even function.

Equation (3.13) immediately suggests other periodic relaxation functions. Two of these are

$$y = \text{Sin}^{-1}\sin x \tag{3.14}$$

$$\rightarrow\rightarrow\rightarrow \quad \text{saw}_\pi x = \text{Tan}^{-1}\tan x \tag{3.15}$$

The first of these is the *sine-derived triangular wave* graphed in Fig. 3.5. It can be written in terms of tri as $\text{tri}(x + \frac{1}{2}\pi) - \frac{1}{2}\pi$. The second, $\text{Tan}^{-1}\tan$

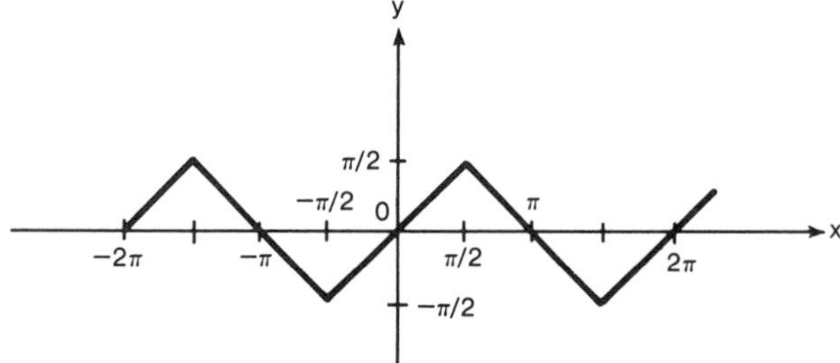

Figure 3.5 Sine-derived triangular wave.

x, is important enough to deserve a special symbol, $\mathrm{saw}_\pi x$. It was briefly considered in Chapter 1. It is the *tangent-derived linear sawtooth wave* graphed in Fig. 3.6. Its values at the discontinuities are shown in Fig. 3.7 where the points at the discontinuities are emphasized by dots.

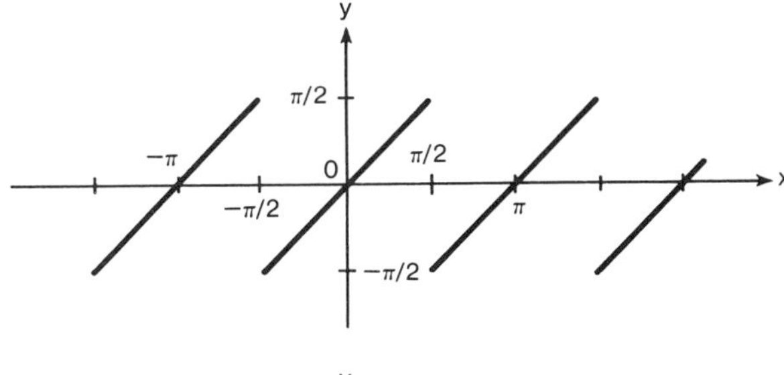

Figure 3.6 Tangent-derived sawtooth wave.

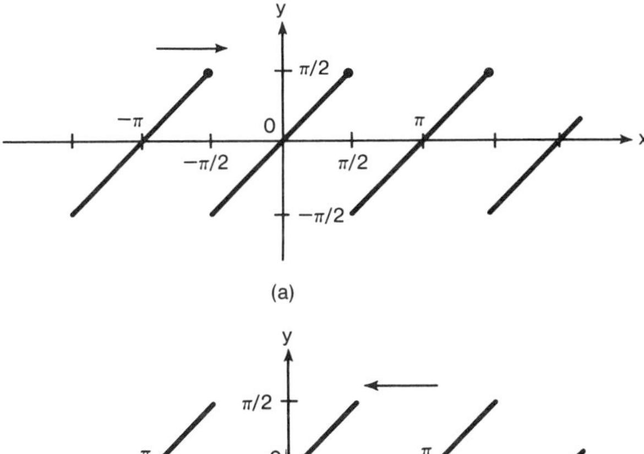

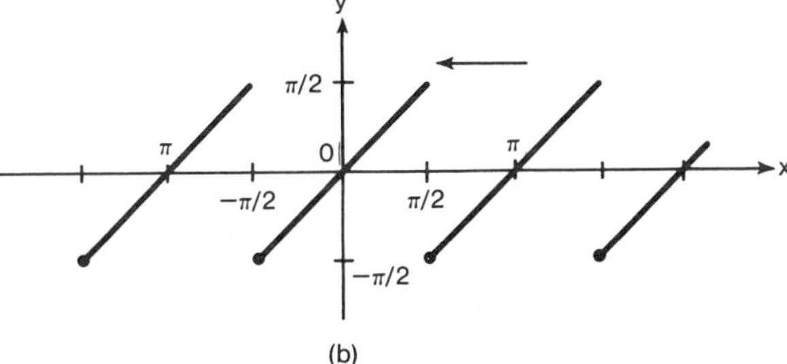

Figure 3.7 Sawtooth wave: The points at the discontinuities.

Proof that infinitesimal hysteresis operates for $\mathrm{saw}_\pi x$ consists of evaluating $\mathrm{Tan}^{-1}\tan x$ at a point of discontinuity by the method of limits used in connection with the signum function. We write for the point at $x = \frac{1}{2}\pi$

$$\mathrm{saw}_\pi(\tfrac{1}{2}\pi) = \lim_{x\to\pi/2} \mathrm{Tan}^{-1}\tan x \tag{3.16}$$

For positive progression the limit is

$$\lim_{x\to\pi/2-} \mathrm{Tan}^{-1}\tan x = \lim_{x\to\pi/2-} x = \tfrac{1}{2}\pi \tag{3.17}$$

because $\mathrm{Tan}^{-1}\tan x = x$ in this cycle. For negative progression the limit is

$$\lim_{x \to \pi/2+} \mathrm{Tan}^{-1}\tan x = \lim_{x \to \pi/2+} (x - \pi) = -\tfrac{1}{2}\pi \qquad (3.18)$$

because $\mathrm{Tan}^{-1}\tan x = x - \pi$ in this cycle. All other points of discontinuity behave similarly.

The function $\mathrm{saw}_\pi x$ has a fundamental period of pi. In general, we write $\mathrm{saw}_a x$ to mean a sawtooth wave of the type of Eq. (3.15) with fundamental period a. That is, $\mathrm{saw}_a x$ is such that

$$\mathrm{saw}_a x = x \qquad \text{when } -\tfrac{1}{2}a \underset{\leftarrow}{\leq} x \underset{\rightarrow}{\leq} \tfrac{1}{2}a \qquad (3.19)$$

and repeats periodically. Its definition is

$$\rightarrow\rightarrow\rightarrow \quad \mathrm{saw}_a x \equiv (a/\pi)\mathrm{Tan}^{-1}\tan(\pi x/a) \qquad (3.20)$$

The notation $\underset{\leftarrow}{\leq} x \underset{\rightarrow}{\leq}$ means $\leqslant x <$ for negative progression and $< x \leqslant$ for positive progression. The function $\mathrm{saw}_a x$ is an odd function with global antisymmetry.

It is easy to show that

$$\mathrm{saw}_a x = (\tfrac{1}{2}a/\pi)\mathrm{saw}_{2\pi}(2\pi x/a) \qquad (3.21)$$

We are now able to write the definition of the regular staircase function $\sigma_a(x)$ of Chapter 2. It is

$$\rightarrow\rightarrow\rightarrow \quad \sigma_a(x) \equiv x - \mathrm{saw}_a x \qquad (3.22)$$

Other functions based on the principal value are legitimate objects of study as well, such as $\mathrm{Tan}^{-1}\mathrm{ctn}\, x$. It will help to investigate some of these. See Appendix F for a listing of functions of this type.

3.4 How to Synthesize Arbitrary Periodic Functions

A useful mathematical property of the sawtooth wave is that it can be used to render periodic the central portion of any function. This property can be stated as follows: If $f(x)$ is any function (nonperiodic or periodic in x with any period), then $f(\mathrm{saw}_a x)$ is periodic in x with period a, with $f(\mathrm{saw}_a x)$ being equal to $f(x)$ for x between $-\tfrac{1}{2}a$ and $\tfrac{1}{2}a$.

This property is formalized in the First Fundamental Theorem.

First Fundamental Theorem

Given: **1.** $p(x)$ is periodic in x with period a

2. $f(x) = p(x)$ in the interval $-\tfrac{1}{2}a \underset{\leftarrow}{\leq} x \underset{\rightarrow}{\leq} \tfrac{1}{2}a$

$\rightarrow\rightarrow\rightarrow \quad p(x) = f(u)$

where $u = \mathrm{saw}_a x$.

Proof of the theorem is given in Appendix A.

Some examples of the use of the First Fundamental Theorem follow.

Example 3.1. $p(x) = \text{tri } x$, $a = 2\pi$, $f(u) = |u|$, giving $\text{tri } x = |\text{saw}_{2\pi}x|$.

Example 3.2. $p(x) = \text{sqr } x$, $a = 2\pi$, $f(u) = \text{sgn } u$, giving $\text{sqr } x = \text{sgn}(\text{saw}_{2\pi}x)$.

Example 3.3. $p(x) = \text{sgn}(\cos x)\sin x$, $a = \pi$, $f(u) = \sin u$, giving $\text{sgn}(\cos x) \sin x = \sin(\text{saw}_{\pi}x)$.

This is the *pseudorectified sinewave* shown in Fig. 3.8. In a similar manner, we define sqr x cos x as the *pseudorectified cosine wave,* shown in Fig. 3.9. The pseudorectified sine and cosine waves appear in the derivative and integral of $|\cos x|$ and $|\sin x|$, respectively, as is shown in Chapters 4 and 5.

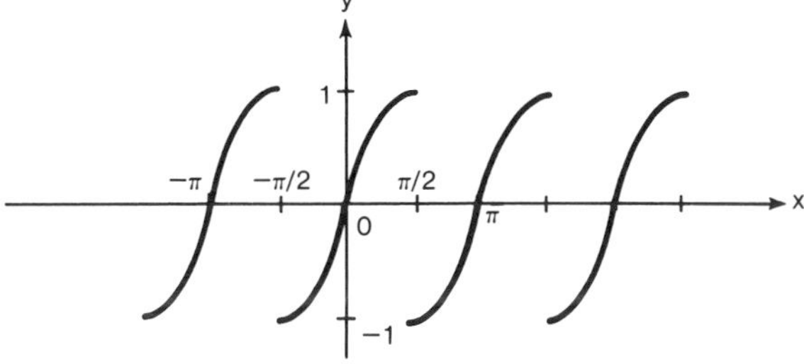

Figure 3.8 Pseudorectified sinewave.

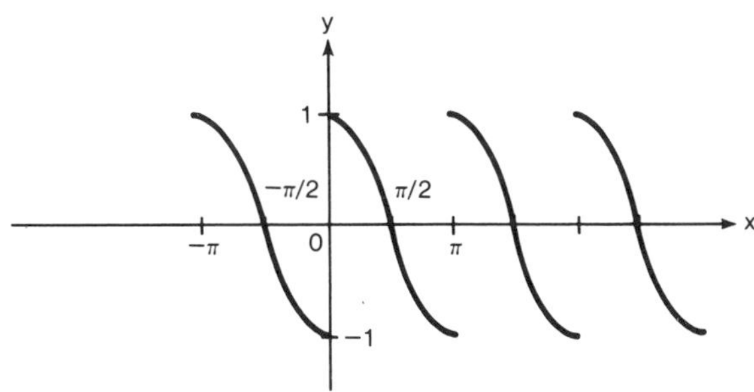

Figure 3.9 Pseudorectified cosinewave.

Example 3.4. $p(x) = |\cos x|$, $a = \pi$, $f(u) = \cos u$, giving $|\cos x| = \cos(\text{saw}_{\pi}x)$.

Example 3.5. $p(x) = \text{saw}_a x$, $a = a$, $f(u) = u$, giving $\text{saw}_a x = \text{saw}_a x$.

This seemingly trivial example illustrates the basic simplicity of the theorem. Subsequent examples illustrate its power.

Example 3.6. $p(x) = \delta(\sin x)$, $a = \pi$, $f(u) = \delta(u)$, giving $\delta(\sin x) = \delta(\text{saw}_{\pi}x)$.

Delta pulses do not occur at the discontinuities in $\text{saw}_{\pi}x$ because $\text{saw}_{\pi}x \neq 0$ there.

Example 3.7. $p(x)$ is the *parabolic wave* shown in Fig. 3.10, $a = \pi$, $f(u) = 1 - (4/\pi^2)u^2$, giving $p(x) = 1 - (4/\pi^2)\ \text{saw}_{\pi}^2 x$.[2]

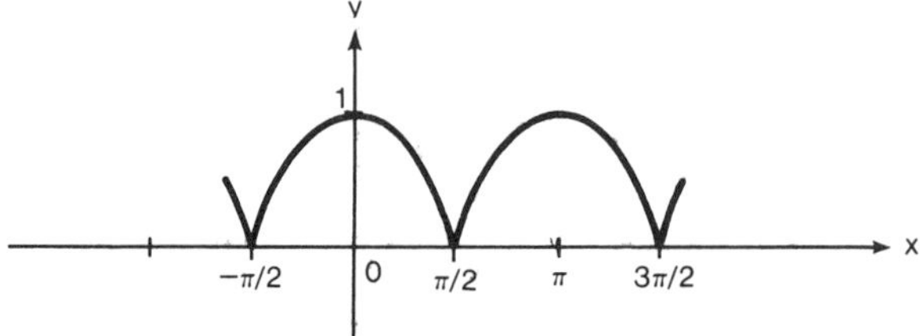

Figure 3.10 Parabolic wave.

Example 3.8. $p(x)$ is as shown in Fig. 3.11, $a = 2\pi, f(u) = S(u)(u/\pi - 1) + 1$, giving $p(x) = S(\text{saw}_{2\pi}x)\left(\frac{1}{\pi}\text{saw}_{2\pi}x - 1\right) + 1 = S(\sin x)\left(\frac{1}{\pi}\text{saw}_{2\pi}x - 1\right) + 1$.

Examples 3.7 and 3.8 show how arbitrary periodic functions can be synthesized.

Other examples useful in the applications follow.

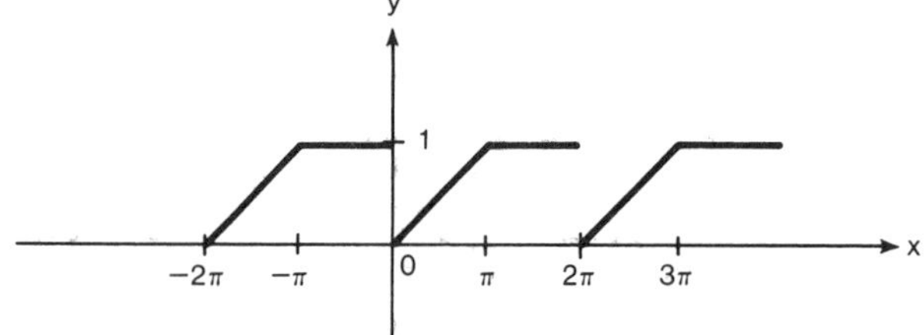

Figure 3.11 The synthesized wave of Example 3.8.

Example 3.9. $p(x)$ is the *semi-interrupted cosinewave* shown in Fig. 3.12, $a = 2\pi$, $f(u) = \cos\frac{\pi u}{\pi + \alpha}$, giving $p(x) = \cos\left(\frac{\pi}{\pi + \alpha}\text{saw}_{2\pi}x\right)$, where α is the *interruption*.

An obvious corollary of the First Fundamental Theorem is:

Corollary 1

Given: $p(x)$ is periodic in x with period a.

$\rightarrow\rightarrow\rightarrow \quad p(\text{saw}_a x) = p(x)$

For example, $\sin(\text{saw}_{2\pi}x) = \sin x$.

[2]We use $\text{saw}_{\pi}^2 x$ to mean $(\text{saw}_{\pi}x)^2$ in agreement with the convention $\sin^2 x \equiv (\sin x)^2$, $\cos^2 x \equiv (\cos x)^2$, etc.

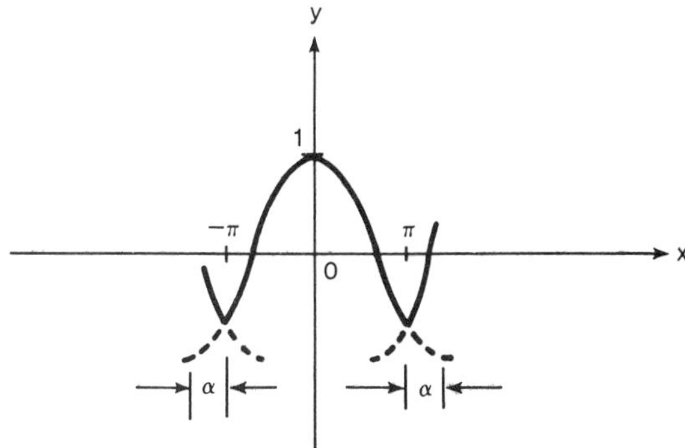

Figure 3.12 Semi-interrupted cosinewave.

Another less obvious corollary is:

Corollary 2

Given: $p(x)$ is an even function periodic in x with period a.

$\rightarrow\rightarrow\rightarrow \quad p(|\text{saw}_a x|) = p(x)$

For example, $\cos(\text{tri } x) = \cos x$.

Thus the function of Example 3.9 can also be written

$$p(x) = \cos\left(\frac{\pi}{\pi + \alpha}\text{tri } x\right) \tag{3.23}$$

We use Corollary 2 to find expressions for the *symmetrically-interrupted cosinewave* of Fig. 3.13 and the *asymmetrically-interrupted cosinewave* of Fig. 3.14 in the following two examples.

Example 3.10. $p(x)$ is the symmetrically-interrupted cosinewave shown in Fig. 3.13, $a = 2\pi$. Between $x = 0$ and $x = \pi$, $p(x)$ is given by

$$y = \cos\frac{\pi(x + \alpha)}{\pi + 2\alpha}, \qquad \text{so that}$$

$$p(x) = \cos\frac{\pi(\alpha + \text{tri } x)}{\pi + 2\alpha}.$$

Example 3.11. $p(x)$ is the asymmetrically-interrupted cosinewave shown in Fig. 3.14, $a = 2\pi$. Between $x = 0$ and $x = \pi$, $p(x)$ is given by

$$y = \cos\frac{\pi(x + \beta)}{\pi + \alpha + \beta}, \qquad \text{so that}$$

$$p(x) = \cos\frac{\pi(\beta + \text{tri } x)}{\pi + \alpha + \beta}.$$

When using $f(|\text{saw}_a x|)$ to synthesize a periodic function from $f(x)$, only that part of $f(x)$ corresponding to the range of x-values from zero to $\frac{1}{2}a$

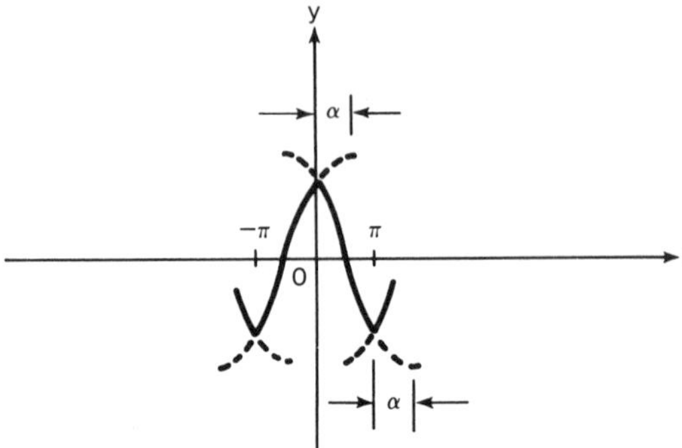

Figure 3.13 Symmetrically interrupted cosinewave.

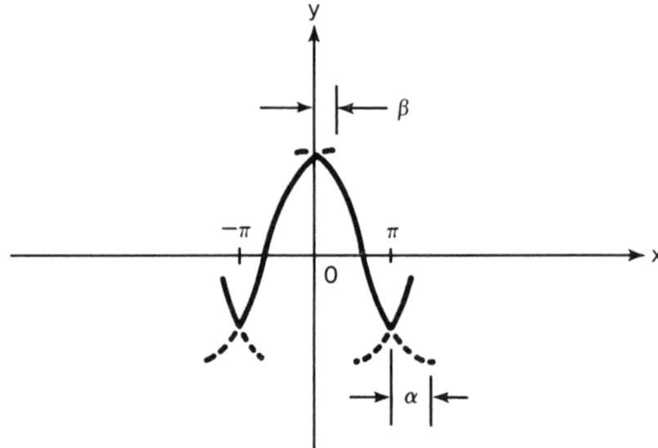

Figure 3.14 Asymmetrically interrupted cosinewave.

is preserved and rendered periodic, because that is the total range of $|\text{saw}_a x|$.

Examples involving periodic staircase functions are given next.

Example 3.12. $p(x)$ is a periodic staircase function, one cycle of which is shown in Fig. 3.15. It obviously is

$$p(x) = \sigma_{\pi/2}(\text{saw}_{2\pi} x).$$

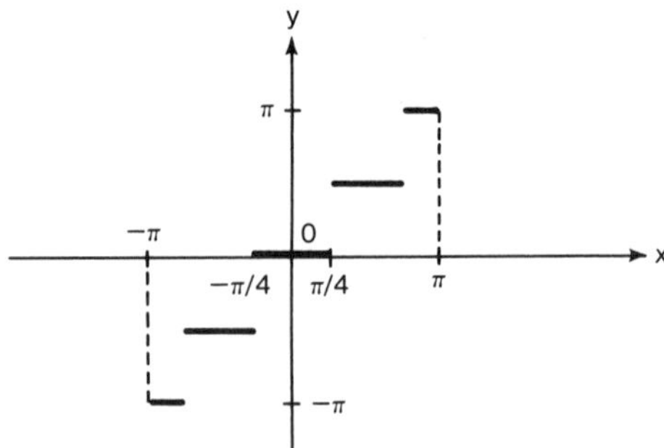

Figure 3.15 A periodic staircase function.

Example 3.13. $p(x)$ is a periodic staircase function, one cycle of which is shown in Fig. 3.16. It is

$$p(x) = \sigma_{\pi/2}(\text{saw}_{2\pi}x - \tfrac{1}{4}\pi).$$

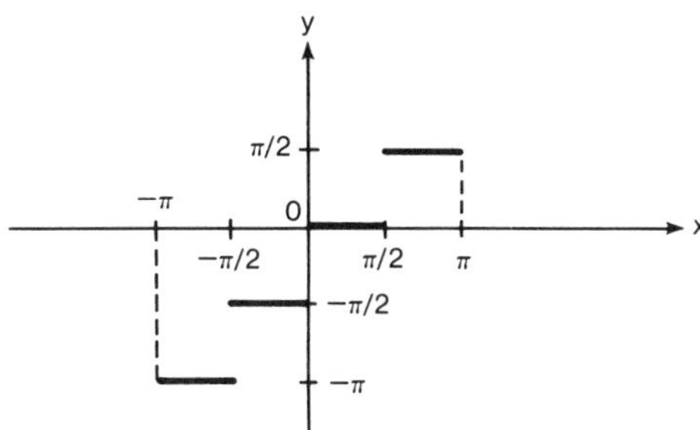

Figure 3.16 A second periodic staircase function.

Example 3.14. $p(x)$ is a periodic staircase function, one cycle of which is shown in Fig. 3.17. It is

$$p(x) = \sigma_{\pi/2}(\text{saw}_{2\pi}x - \tfrac{1}{4}\pi) + \tfrac{1}{4}\pi.$$

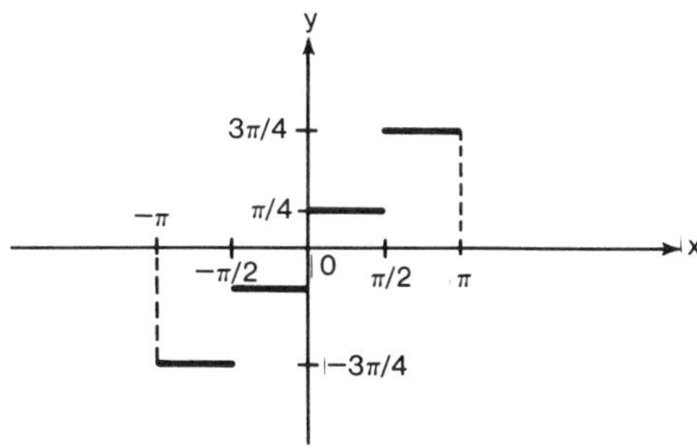

Figure 3.17 A third periodic staircase function.

In example 3.12, the number of steps is found by dividing the index of *saw* by the index of σ. In Examples 3.13 and 3.14, the number of steps is one less than this. From these relationships it is easy to see how to synthesize periodic staircase functions with any desired number of steps.

3.5 Frequency Doubling

We noted in Chapter 2 that the fundamental frequency of δ (sin x) is twice that of sin x. Thus the Dirac delta function acts as a *frequency-doubling operator* in this case. That is because it senses zero crossings, but does not distinguish positive-to-negative crossings from negative-to-positive ones. Thus there is a loss of information as well as a change of waveform. Other cases of frequency doubling have been encountered in the present chapter. These and others are considered next.

An important class of frequency-doubling operations can be characterized as follows. If $p(x)$ and $q(x)$ are periodic in x with fundamental period a and *antiperiodic*[3], then $p'(x) = p(x)q(x)$ is periodic in x with period $\frac{1}{2}a$ (or is a constant). This is easy to prove, for

$$p(x + \tfrac{1}{2}a)q(x + \tfrac{1}{2}a) = [-p(x)][-q(x)] = p(x)q(x) \tag{3.24}$$

or

$$p'(x + \tfrac{1}{2}a) = p'(x) \tag{3.25}$$

which is the condition that $p'(x)$ be periodic in x with period $\frac{1}{2}a$.

Examples of antiperiodic functions are $\sin x$, $\cos x$, sqr x, sgn($\cos x$), tri $x - \frac{1}{2}\pi$, and $\text{Sin}^{-1}\sin x$. These all have a fundamental period of 2π. The product of any two has a fundamental period of π (or is a constant, e.g., $\text{sqr}^2 x = 1$). Thus $\sin^2 x$ has a fundamental period of π as has sqr x $\sin x = |\sin x|$.

A frequency-doubling operation may or may not preserve waveform. In general it does not. Thus $\sin^2 x$ and $\cos^2 x$ are sinusoidal, but sqr $x \sin x$ is neither sinusoidal nor squarewave in form. Even when waveform is preserved, frequency doubling results in a loss of information. This is due to the fact that adjacent half cycles of the original wave may or may not differ from each other, whereas the corresponding portions of the resultant wave are always alike. For example, $\sin^2 x$ can result from the product of $\sin x$ and $\sin x$, but it can also result from the product of $-\sin x$ and $-\sin x$, from the product of $|\sin x|$ and $|\sin x|$, and from the product of $-|\sin x|$ and $-|\sin x|$. Another way of stating this property (rather than as a "loss of information") is to say that frequency-doubling operations *do not have a unique inverse* in general.

Thus there is no way of knowing in the expression $p(x) = \pm(\sin^2 x)^{1/2}$, for example, which sign to use unless we further specify what we want $p(x)$ to be. Such is the situation with trigonometric half-angle identities such as $p(x) = \pm[(1-\cos x)/2]^{1/2}$. If we want $p(x)$ to be $\sin \frac{1}{2}x$ (as opposed to, say, $|\sin\frac{1}{2}x|$), then we can write $\sin \frac{1}{2}x = \text{sqr}\ \frac{1}{2}x[(1 - \cos x)/2]^{1/2}$; but in order to determine the function sqr $\frac{1}{2}x$, we must know $\sin \frac{1}{2}$x or x beforehand!

If the fundamental periods of $p(x)$ and $q(x)$ are both a and the fundamental period of $p'(x) = p(x)q(x)$ is $\frac{1}{2}a$, then we can write

$$p(\text{saw}_a x)q(\text{saw}_a x) = p(\text{saw}_{a/2} x)q(\text{saw}_{a/2} x) \tag{3.26}$$

by Corollary 1 to the First Fundamental Theorem. For example,

$$\text{sqr}\ x \sin x = \text{sqr}\ 2x \sin(\text{saw}_\pi x) \tag{3.27}$$

[3] I.e., $p(x + \frac{1}{2}a) = -p(x)$ and $q(x + \frac{1}{2}a) = -q(x)$.

because

$$\begin{aligned} \text{sqr}\, x \sin x &= \text{sqr}(\text{saw}_{2\pi}x) \sin(\text{saw}_{2\pi}x) \\ &= \text{sqr}(\text{saw}_{\pi}x) \sin(\text{saw}_{\pi}x) \\ &= \text{sqr}\, 2x \sin(\text{saw}_{\pi}x) \end{aligned}$$

This can easily be confirmed by graphing.

The value of sgn(cos x) between $-\frac{1}{2}\pi$ and $\frac{1}{2}\pi$ is $+1$. Therefore $\text{sgn}[\cos(\text{saw}_{\pi}x)] = 1$. This fact, coupled with the frequency-doubling properties just described, can be used to find a particular class of identities defined by

$$\text{sgn}(\cos x)q(x) = q(\text{saw}_{\pi}x) \tag{3.28}$$

where q(x) is antiperiodic with period 2π.

The identities $\text{sgn}(\cos x) \sin x = \sin(\text{saw}_{\pi}x)$ and $\text{sgn}(\cos x)\cos x = \cos(\text{saw}_{\pi}x)$ of Examples 3.3 and 3.4 belong to this class. Other examples are

$$\text{sgn}(\cos x)(\text{tri}\, x - \tfrac{1}{2}\pi) = \text{tri}(\text{saw}_{\pi}x) - \tfrac{1}{2}\pi \tag{3.29}$$

which is graphed in Fig. 3.18;

$$\text{sgn}(\cos x)\text{Sin}^{-1}\sin x = \text{Sin}^{-1}\sin(\text{saw}_{\pi}x) = \text{saw}_{\pi}x \tag{3.30}$$

$$\text{sgn}(\cos x)\text{sqr}\, x = \text{sqr}\, 2x \tag{3.31}$$

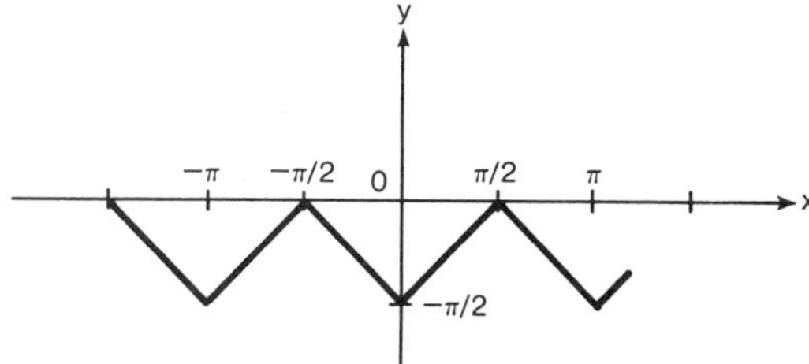

Figure 3.18 The triangular wave of Eq. (3.29).

Since $[\text{sgn}(\cos x)]^{-1} = \text{sgn}(\cos x)$, we have the additional relationships from the foregoing ones:

$$\text{sgn}(\cos x)\sin(\text{saw}_{\pi}x) = \sin x \tag{3.32}$$

$$\text{sgn}(\cos x)\cos(\text{saw}_{\pi}x) = \cos x \tag{3.33}$$

$$\text{sgn}(\cos x)[\text{tri}(\text{saw}_{\pi}x) - \tfrac{1}{2}\pi] = \text{tri}\, x - \tfrac{1}{2}\pi \tag{3.34}$$

$$\text{sgn}(\cos x)\text{saw}_{\pi}x = \text{Sin}^{-1}\sin x \tag{3.35}$$

$$\text{sgn}(\cos x)\text{sqr}\, 2x = \text{sqr}\, x \tag{3.36}$$

These can all easily be confirmed by graphing.

Other frequency-doubling expressions that we will find useful can be found as follows. Let $p'(x) = \text{sqr } x(\text{tri } x - \frac{1}{2}\pi) = \text{saw}_{2\pi}x - \frac{1}{2}\pi \text{ sqr } x$. The fundamental periods of $\text{saw}_{2\pi}x$ and sqr x are both 2π; but $p'(x)$ has the fundamental period π, giving finally

$$\text{saw}_{2\pi}x - \tfrac{1}{2}\pi \text{ sqr } x = \text{saw}_{\pi}x - \tfrac{1}{2}\pi \text{ sqr } 2x \tag{3.37}$$

This can easily be confirmed by graphing. A variation of Eq. (3.37) is

$$\text{sqr } 2x + \frac{2}{\pi}\sigma_{\pi}(x) = \text{sqr } x + \frac{2}{\pi}\sigma_{2\pi}(x) \tag{3.38}$$

We began this section by stating that $\delta(\sin x)$ is a frequency-doubling expression. This can be interpreted in terms of the frequency-doubling operation just discussed; for $\delta(u)$ is of the form $p'(x) = p(x)q(x)$ where $p(x) = [2\ du/dx]^{-1}$ and $q(x) = d(\text{sgn } u)/dx$. If we let $u = \sin x$, it is easy to see that both p and q are antiperiodic with period 2π.

3.6 Transfer Diagrams and Function Fitting

Throughout this chapter we have encountered expressions of the form

$$f(x) = g(u) \tag{3.39}$$

where $u = u(x)$. For example, Eq. (3.13) is in this form where $f(x) =$ tri x, $g(u) = \text{Cos}^{-1}u$, and $u = \cos x$. Equation (3.39) can be represented by the block diagram in Fig. 3.19, where g is the *transfer characteristic* of the *function generator*, $u(x)$ is the *input function*, and $f(x)$ is the *output function*.

This situation can also be represented by a *transfer diagram* such as that shown in Fig. 3.20, where we have used Eq. (3.13) as an example.

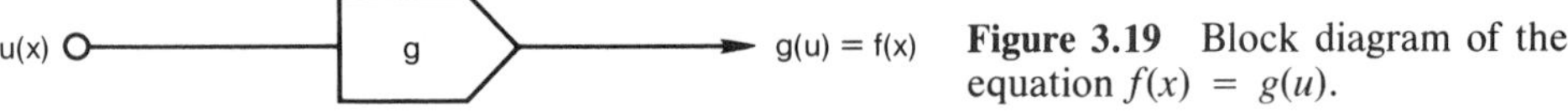

Figure 3.19 Block diagram of the equation $f(x) = g(u)$.

In the transfer diagram, each point of the output function is determined by projecting from a point of the input function vertically to the transfer characteristic, then horizontally to the corresponding x-coordinate of the output function.[4]

Other functions can be used as inputs to a particular function generator to get corresponding outputs. For example, it can readily be shown by means of a transfer diagram that $\text{Cos}^{-1}\sin x$ is a triangular wave of the form tri x, but "pi-halves out of phase" with it. That is,

An alternate form of transfer diagram is illustrated in Appendix F.

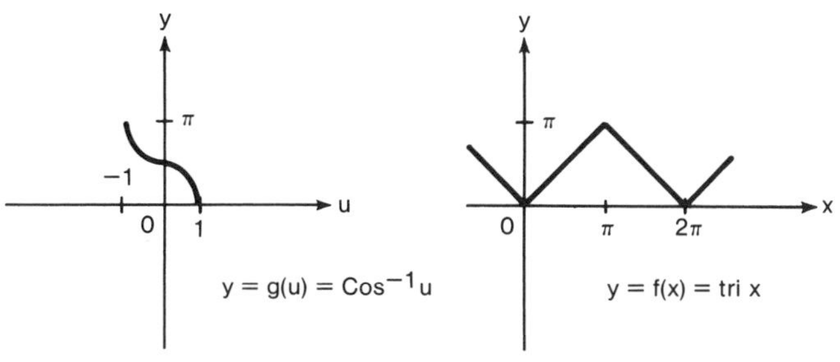

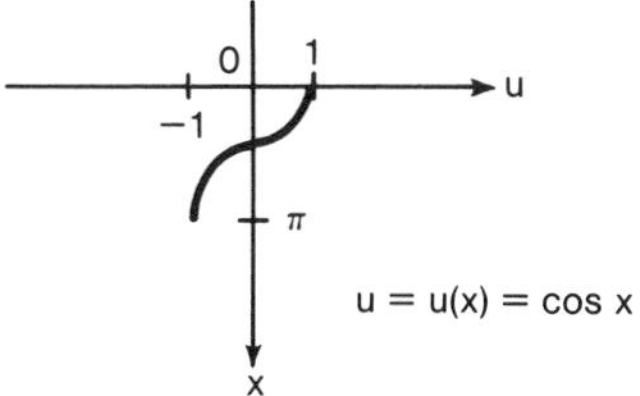

Figure 3.20 A transfer diagram.

$$\text{Cos}^{-1}\sin x = \text{tri}(x + \tfrac{1}{2}\pi) \tag{3.40}$$

As another example transfer diagram, consider the equation of Example 3.4,

$$|\cos x| = \cos(\text{saw}_\pi x) \tag{3.41}$$

The transfer diagram for this case is shown in Fig. 3.21.

As a third example we consider the following case. A sine function generator can be used to produce sinewaves from triangular waves. This is expressed as

$$\sin x = \sin(\text{Sin}^{-1}\sin x) \tag{3.42}$$

where $\text{Sin}^{-1}\sin x$ is the sine-derived triangular wave graphed in Fig. 3.5. This is illustrated in the transfer diagram shown in Fig. 3.22.

This method of producing sinewaves from triangular waves is currently popular in integrated circuit function generators. In this application, the sine transfer characteristic does not stop abruptly at $-\frac{1}{2}\pi$ and $+\frac{1}{2}\pi$ but continues on in some manner. It is apparent that such a function generator produces a waveform whose shape is highly dependent on the amplitude and d-c component of the input triangular wave—as well as on the shape of that triangular wave. Also, accuracy is dependent on the accuracy of the transfer characteristic.

The technique of generating electronic embodiments of functions by the use of transfer characteristics, as just described, is called *function fitting*.

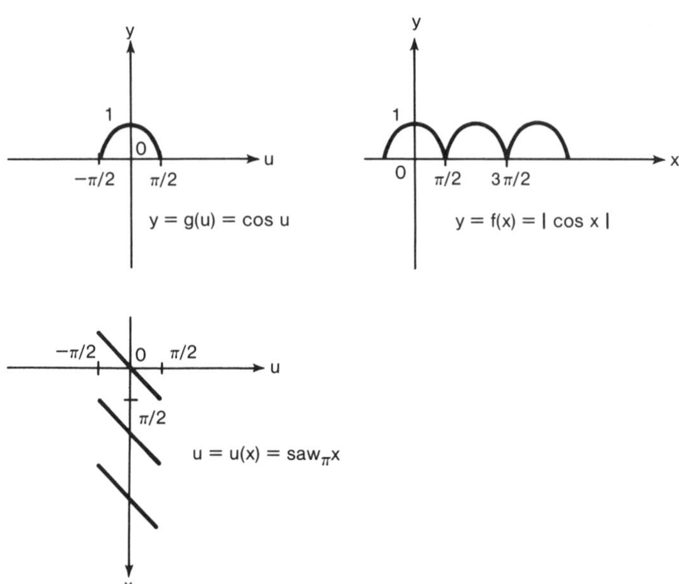

Figure 3.21 Transfer diagram of Eq. (3.41).

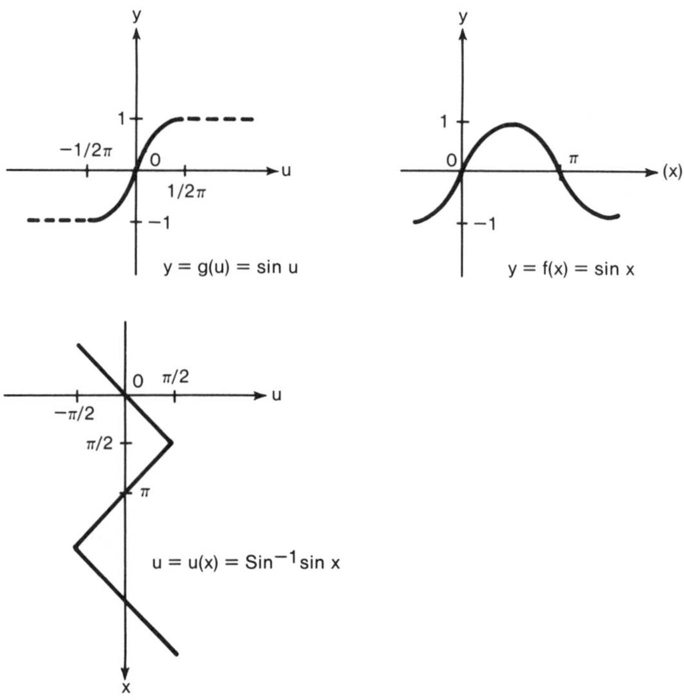

Figure 3.22 Transfer diagram showing how a sinewave can be produced from a triangular wave by use of a sinusoidal transfer characteristic.

Finally, the First Fundamental Theorem and its corollaries can be illustrated by the use of transfer diagrams. This is shown in Figs. 3.23 and 3.24. Figure 3.23 illustrates the Theorem and Corollary 1. Figure 3.24 illustrates Corollary 2.

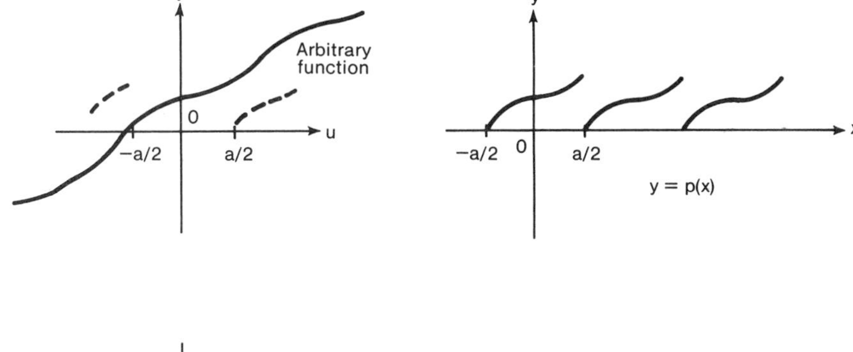

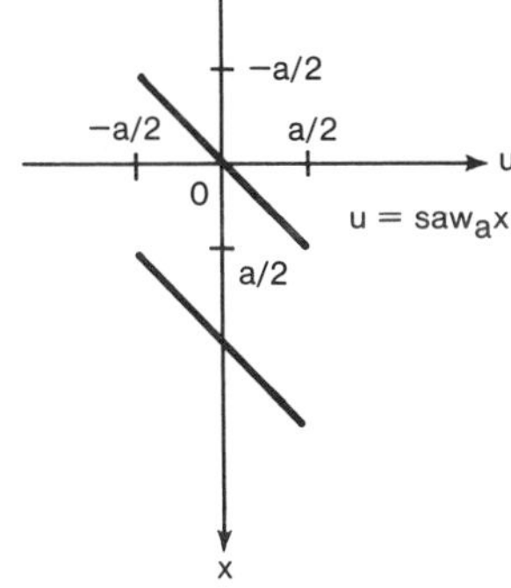

Figure 3.23 Transfer diagram illustrating the First Fundamental Theorem and Corollary 1.

3.7 Summary

The methods discussed in this chapter for generating nonsinusoidal periodic functions are summarized in the following *principle of periodicity* and lemma.

Principle of Periodicity. If $f(x)$ is a single-valued function of x and $f(u)$ is periodic in x where $u = u(x)$, then the *principle of periodicity* is said to hold for $f(u)$.

Lemma 3.1

Given: **1.** The function $f(x)$ is a single-valued function of x.

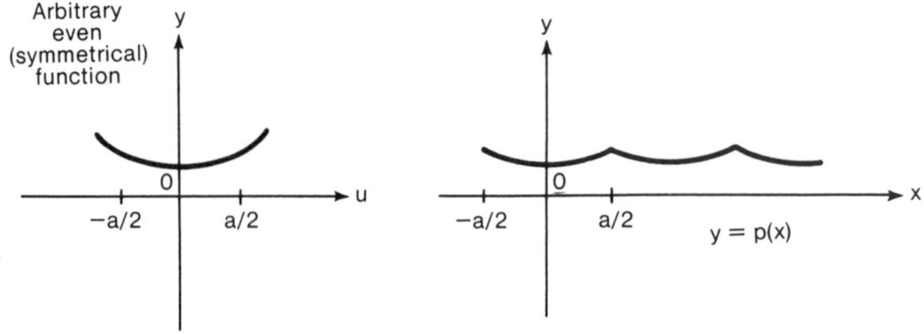

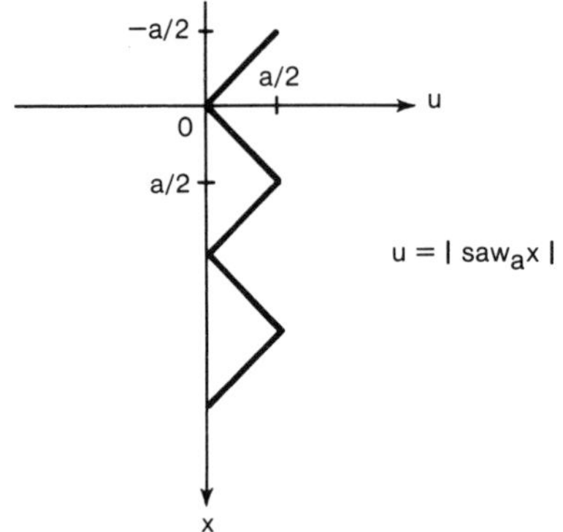

Figure 3.24 Transfer diagram illustrating Corollary 2.

2. The operation $f()$ can be represented as a transfer characteristic. →→→ It is sufficient that $u = u(x)$ be periodic in x in order for the principle of periodicity to hold for $f(u)$. Further, the fundamental period of u is also a period of $f(u)$.

Thus $|u|$, sgn u, $\mathrm{Cos}^{-1}u$, $\mathrm{Sin}^{-1}u$, and $\mathrm{Tan}^{-1}u$ are all periodic when u is periodic. Also, $f(u)$ is periodic when $u = \mathrm{saw}_a x$. Cf. the First Fundamental Theorem.

Lemma 3.1 is presented as a summary of material in this chapter, so its general proof is omitted. *Given 2* of the lemma serves to exclude the operations of differentiation and integration. These operations (they are both linear operations) are the subjects of the following two chapters. Cf. Lemmas 4.1 and 5.1 for counterparts of Lemma 3.1.

4

HOW TO DIFFERENTIATE RELAXATION FUNCTIONS

4.1 The Derivative of a Broken Function Is a Discontinuous Function

We know from elementary differential calculus that an integrable function has no unique integral. For example, $\sin x$ is an integral of $\cos x$; but $\sin x + 1$, $\sin x + 2$, etc., are also integrals of $\cos x$. That is, we can add an *arbitrary constant of integration* because the derivative of a constant is zero.

Similarly, it is shown in works on the integral calculus and the Laplace transform that a differentiable function has no unique derivative. For example, $\cos x$ is a derivative of $\sin x$; but $\cos x + N(x)$ is also a derivative of $\sin x$, where $N(x)$ is a so-called *null function* that may not be literally zero, but that integrates like zero. $N(x)$ is an *arbitrary null function of differentiation*. The vestigial delta function $|x|\delta(x)$ is one such "function"; $x\delta(x)$ is another. We will occasionally encounter expressions of this type when differentiating.

In practice, we can choose either to add an arbitrary constant to integrals or an arbitrary null function to derivatives; we need not do both. It is usual practice to do the former. In keeping with this practice, in the present volume we retain the constant of integration and discard the null function of differentiation.

Continuous and discontinuous functions require different considerations with regard to differentiation. In Chapter 2, the delta function was introduced to use in formulating differentials of step and staircase functions. In general, functions that have one or more finite discontinuities contain delta functions in their differentials, whereas those that are not discontinuous do not require them.

The derivatives of continuous-smooth functions are defined in elementary calculus. Those definitions also apply to continuous-smooth portions of continuous-broken and discontinuous functions.

The derivative of a continuous function $y = F(x)$ at a point that is not a breakpoint has a unique value. By contrast, there are *two* derivatives at a breakpoint at $x = a$, depending on the manner in which $\Delta y/\Delta x$ approaches the limit. Widder[1] called these the "derivative on the left" and the "derivative on the right," $f_-(a)$ and $f_+(a)$ respectively, where

$$\rightarrow\rightarrow\rightarrow \left\{ \begin{aligned} f_-(a) &\equiv \lim_{\Delta x \to 0-} \frac{F(a + \Delta x) - F(a)}{\Delta x} \\ f_+(a) &\equiv \lim_{\Delta x \to 0+} \frac{F(a + \Delta x) - F(a)}{\Delta x} \end{aligned} \right\} \tag{4.1}$$

We will call them the *derivative for positive progression* and the *derivative for negative progression,* respectively.[2]

The function $F(x)$ is said to be *normally differentiable* at $x = a$ if

$$f(a \pm 0) = f_\pm(a) \tag{4.2}$$

where $f_\pm(a)$ is as defined in Eqs. (4.1) and

$$f(a \pm 0) = \lim_{x \to a\pm} \frac{dF(x)}{dx} \tag{4.3}$$

This means that its derivative at the point can be found in the usual manner, i.e., by applying the elementary differentiation formulas then taking the unilateral limits $x \to a\pm$.

Table 4.1 contains some useful differentiation formulas. Others can be found in standard tables of derivatives. Other derivatives that cannot be found in standard tables of derivatives are developed throughout this chapter. These are collected together in Appendix H.

As an example we consider $F(x) = |x|$. It can be shown that its derivative is

$$\frac{d}{dx}|x| = \operatorname{sgn} x \tag{4.4}$$

[1]D. V. Widder, *Advanced Calculus* (Englewood Cliffs, N.J.: Prentice-Hall, Inc., 1947), p. 6.

[2]In the present volume, upper and lower case letters in a given equation are used to refer generally to a function and its derivative respectively, or to a function and its integral anti-respectively.

TABLE 4.1 Derivatives

The derivative of u with respect to x, $\frac{d}{dx}u$, is also the ratio of the differentials du and dx. That is, $\frac{d}{dx}u = \frac{du}{dx} = du/dx$ and $dx/dx = 1$. We also write $(d/dx)u$ for $\frac{d}{dx}u$.

GENERAL FORMS

1. Derivative of a sum

$$\frac{d}{dx}(u + v) = du/dx + dv/dx$$

2. Derivative of a product

$$\frac{d}{dx}(uv) = u\,dv/dx + v\,du/dx$$

3. Derivative of a quotient

$$\frac{d}{dx}\frac{u}{v} = \frac{v\,du/dx - u\,dv/dx}{v^2}, \qquad v \neq 0$$

4. Derivative of a "parallel" combination, $u \| v = \dfrac{uv}{u + v}$,

$$\frac{d}{dx}(u\|v) = \frac{u^2 dv/dx + v^2 du/dx}{(u + v)^2}, \qquad u + v \neq 0$$

5. Derivative of a function of a function

$$\frac{d}{dx}f(u) = \frac{d}{du}f(u)\frac{du}{dx}$$

6. Reciprocal of a derivative

$$\frac{1}{du/dx} = dx/du, \qquad du/dx \neq 0$$

SPECIFIC FORMS

7. $\frac{d}{dx}a = 0,\qquad a = \text{constant}$

8. $\frac{d}{dx}i = 0,\qquad i^2 = -1$

9. $\frac{d}{dx}x^2 = 2x$

10. $\frac{d}{dx}\sin x = \cos x$

11. $\frac{d}{dx}\cos x = -\sin x$

12. $\frac{d}{dx}e^x = e^x,\qquad e = 2.71828...$

As proof, we write $u = |x|$, recognize that

$$u^2 = x^2 \tag{4.5}$$

and differentiate both sides using Forms 5 and 9 from Table 4.1, giving

$$2u\frac{d}{dx}u = 2x \tag{4.6}$$

Equation (4.6) is readily solved to give the final result in Eq. (4.4). It is easy to show that Eq. (4.4) meets the test of Eqs. (4.1) for $x = a = 0$ because

$$\lim_{\Delta x \to 0\pm} \frac{|0 + \Delta x| - |0|}{\Delta x} = \pm 1 \tag{4.7}$$

Similarly, it can be shown that

$$\frac{d}{dx}|u| = \operatorname{sgn} u \frac{du}{dx} \tag{4.8}$$

by starting with the identity[3] $|u|^2 = u^2$ and differentiating both sides. As an example, we let $u = \sin x$, giving

$$\frac{d}{dx}|\sin x| = \operatorname{sqr} x \cos x \tag{4.9}$$

By starting with $|\sin u|^2 = \sin^2 u$ it can be shown that

$$\frac{d}{dx}|\sin u| = \operatorname{sqr} u \cos u \frac{du}{dx} \tag{4.10}$$

and by starting with $|\cos u|^2 = \cos^2 u$ it can be shown that

$$\frac{d}{dx}|\cos u| = \operatorname{sgn}(\cos u)\,(-\sin u)\frac{du}{dx} \tag{4.11}$$

4.2 The Derivative of a Discontinuous Function Contains a Delta Function

It is said in standard analysis that the derivative at a discontinuity does not exist. The fact is, the classical definition of derivative as the limit of $\Delta y/\Delta x$ does not readily lend itself to finding the derivative at a discontinuity. For this case we need an auxiliary definition of derivative.

Equation (2.30) can form the basis for such a definition. Accordingly, we define the derivative of the Heaviside step function as

$$\rightarrow\rightarrow\rightarrow \quad \frac{d}{dx}S(u) \equiv \delta(u)\frac{du}{dx} \tag{4.12}$$

[3] Or by use of Form 5 of Table 4.1.

It follows that the derivative of the signum function is

$$\frac{d}{dx}\,\text{sgn}\,u = 2\delta(u)\frac{du}{dx} \tag{4.13}$$

(More is said about the relationship between dS/dx and the limit of $\Delta S/\Delta x$ in Appendix B.) Mathews and Walker[4] are so bold as to state without equivocation that $(d/dt)S(t) = \delta(t)$. We restrict $u = u(x)$ in Eqs. (4.12) and (4.13) to continuous functions in the present volume so that du/dx itself will not contain delta functions. This is not a serious restriction; for a continuous function v can always be found such that sgn v = sgn u when u is discontinuous.

In general, the derivative of a discontinuous function $F(x)$ with a single finite discontinuity at $x = a$ is defined as follows. Let the amount of the discontinuity be r [cf. Eq. (1.7)], where

$$r = F(a + 0) - F(a - 0) \tag{4.14}$$

with $F(a + 0)$ and $F(a - 0)$ being convenient notations for

$$F(a + 0) = \lim_{x\to a+} F(x) \tag{4.15a}$$

$$F(a - 0) = \lim_{x\to a-} F(x) \tag{4.15b}$$

First we define a function $G(x)$ that is continuous at $x = a$ (and everywhere else) as

$$G(x) = F(x) - rS(x - a) \tag{4.16}$$

$G(x)$ is called the "shape" of $F(x)$ because it has the same structure as $F(x)$ except that the discontinuity has been removed. The derivative of $F(x)$ is now defined as

$$\rightarrow\rightarrow\rightarrow \quad \frac{d}{dx}F(x) = \frac{d}{dx}G(x) + r\delta(x - a) \tag{4.17}$$

It is clear that $dG(x)/dx$ is the kind of derivative discussed in the previous section.

Derivatives of multiply-discontinuous functions are similarly defined. For example, in Chapter 2 we saw that $\pi\int\delta(\cos x)\,dx = \sigma_\pi(\text{x}) + C$. It follows that

$$\frac{d}{dx}\sigma_\pi(x) = \pi\delta(\cos x) \tag{4.18}$$

[4]J. Mathews and R. L. Walker, *Mathematical Methods of Physics,* 2nd ed. (New York: W. A. Benjamin, Inc., 1970), p. 107.

Since $\sigma_\pi(x) = x - \text{saw}_\pi x$, it further follows that[5]

$$\frac{d}{dx}\text{saw}_\pi x = 1 - \pi\delta(\cos x) \tag{4.19}$$

These are both multiply discontinuous functions whose derivatives presented above conform to the definition in Eq. (4.17) throughout, including all points at which discontinuities occur.

As another example we find the derivative of the sine-derived squarewave using Eq. (4.13). We write

$$\frac{d}{dx}\text{sqr}\, x = \frac{d}{dx}\,\text{sgn}(\sin x) = 2\,\delta(\sin x)\cos x \tag{4.20}$$

This is the sine-derived alternating delta pulse train introduced in Chapter 2.

We consider here only derivatives of finitely-discontinuous (piecewise continuous) functions. Derivatives of infinitely-discontinuous functions are considered in Chapter 9.

The derivative of a finitely-discontinuous function thus consists of two parts, a part not containing delta pulses indicative of the shape of the function without regard to discontinuities, and a part consisting entirely of delta pulses indicative of the discontinuity pattern. (In effect, the delta function is a bookkeeping tool for keeping track of the positions, magnitudes, and polarities of discontinuities in a differentiated function.) The first part is the *intrinsic derivative*. The sum of the two parts is the *complete derivative*. We can also speak of the "intrinsic derivative" when referring to continuous functions. In this case the intrinsic derivative is also the complete derivative. If the intrinsic derivative of a finitely-discontinuous function is zero, then the complete derivative consists simply of a delta function. Such is the case for the squarewave as shown in an earlier example.

A better definition of "shape" than the one given above can now be formulated as follows. *If the complete derivative of $G(x)$ is the intrinsic derivative of $F(x)$, then $G(x)$ is called the shape of $F(x)$.* If $G(x)$ is the shape of $F(x)$, it follows that $G(x) + C$ is also the shape of $F(x)$.

All derivatives in this volume are complete derivatives unless otherwise noted.

Examples of the use of Eq. (4.12) follow.

[5]This and the previous differentiation are presented here without proof. They are proved in the following section. Cf. Example 4.3.

Example. The derivative of $xS(x)$ is

$$\frac{d}{dx}xS(x) = S(x) + x\delta(x) = S(x)$$

because $x\delta(x)$ is a null function. We note that $xS(x)$ is broken and its derivative, $S(x)$, is discontinuous.

Example. The derivative of $x^2S(x)$ is

$$\frac{d}{dx}x^2S(x) = 2xS(x) + x^2\delta(x) = 2xS(x).$$

We note that $x^2S(x)$ is smooth and its derivative, $2xS(x)$, is broken.

The above two functions, $xS(x)$ and $x^2S(x)$, can be differentiated more simply by use of the following theorem.

Theorem 4.1. When differentiating a function that contains no jumps, only the intrinsic portion of intermediate derivatives need be taken.

Proof is beyond the scope of this volume; however, the outline of a proof is given in Appendix A. It rests basically on the fact that the derivative of a continuous function, or one with no jumps, has no delta function component. With the use of this theorem, we see that $S(x)$ differentiates like a constant in the two examples above because its intrinsic derivative is zero. Thus the results, $S(x)$ and $2xS(x)$, are obtained virtually by inspection.

As another example of the use of Theorem 4.1, the problem of Eq. (4.4) is solved by simply writing $(d/dx)\,|x| = (d/dx)x \operatorname{sgn} x = \operatorname{sgn} x$, because sgn x differentiates like a constant in this case.

Constants, step functions, and staircase functions have intrinsic derivatives of zero; therefore they are collectively referred to as *zero-slope functions*. We now present a corollary to Theorem 4.1 that involves zero-slope functions.

Corollary

Given: **1.** stair (x) is a zero-slope function.

2. $F(x)$ + stair(x) is the shape of $F(x)$, i.e., it is continuous, but $F(x)$ need not be.

$$\rightarrow\rightarrow\rightarrow \quad \frac{d}{dx}[F(x) + \text{stair}(x)] = \phi(x)$$

where $\phi(x)$ is the intrinsic derivative of $F(x)$.

Example. Let $F(x) = \operatorname{sqr} x \cos x$. Graphing shows that $F(x)$ + stair(x) is continuous if stair$(x) = -\dfrac{2}{\pi}\sigma_\pi(x - \tfrac{1}{2}\pi)$. Therefore

$$\frac{d}{dx}\left[\text{sqr } x \cos x - \frac{2}{\pi}\sigma_\pi(x - \tfrac{1}{2}\pi)\right] = \text{sqr } x \frac{d}{dx}\cos x = -|\sin x|.$$

Thus we are able to anticipate a result from the following chapter, namely that

$$\int |\sin x|\, dx = -\text{sqr } x \cos x + (2/\pi)\,\sigma_\pi(x - \tfrac{1}{2}\pi) + C.$$

4.3 How to Differentiate Nonsinusoidal Periodic Functions

From standard analysis, the Fourier series representation of the squarewave of Fig. 1.1 is

$$f(x) = \frac{4}{\pi}\sum_{n=1,3,5,\ldots} (1/n)\sin nx \tag{4.21}$$

and that of the triangular wave of Fig. 1.2 is

$$g(x) = \frac{\pi}{2} - \frac{4}{\pi}\sum_{n=1,3,5,\ldots} (1/n^2)\cos nx \tag{4.22}$$

It is easy to show that a term-by-term differentiation of $g(x)$ of Eq. (4.22) produces the function $f(x)$ of Eq. (4.21).

We now consider this problem using the notation of relaxation analysis. The function $g(x)$ is, except possibly for a null function, equal to the function we have previously defined as tri x. To differentiate it we proceed as follows. We write

$$\frac{d}{dx}\text{tri } x = \frac{d}{dx}\text{Cos}^{-1}u \tag{4.23}$$

where $u = \cos x$. This gives

$$\begin{aligned}\frac{d}{dx}\text{tri } x &= \frac{d}{du}\text{Cos}^{-1}u\,\frac{du}{dx}\\ &= -(1 - \cos^2 x)^{-1/2}(-\sin x)\\ &= \frac{\sin x}{|\sin x|} = \text{sqr } x\end{aligned} \tag{4.24}$$

which is the function $f(x)$, except possibly for a null function.

In a similar manner, the sine-derived triangular wave can be differentiated using classical techniques. We write

$$\frac{d}{dx}\text{Sin}^{-1}\sin x = (1 - \sin^2 x)^{-1/2}\cos x = \frac{\cos x}{|\cos x|} = \text{sgn}(\cos x) \tag{4.25}$$

obtaining the cosine-derived squarewave as expected.

Other periodic functions can be differentiated using the formulas of Table 4.1. The derivative of the interrupted cosinewave of Example 3.11 is readily found to be

$$\frac{d}{dx}\cos\frac{\pi(\beta + \operatorname{tri} x)}{\pi + \alpha + \beta} = -\frac{\pi}{\pi + \alpha + \beta}\sin\frac{\pi(\beta + \operatorname{tri} x)}{\pi + \alpha + \beta}\operatorname{sqr} x \tag{4.26}$$

A wide variety of periodic functions can be differentiated by use of the Second Fundamental Theorem presented below. We begin with a self-evident lemma.

Lemma 4.1. If $p(x)$ is periodic in x with period a, then $dp(x)/dx$ is also periodic in x with period a.

As a result of this lemma, if $F(x)$ of the corollary to Theorem 4.1 is periodic, the derivative of $F(x) + \operatorname{stair}(x)$ is also periodic (even though $F(x) + \operatorname{stair}(x)$ may not be periodic), as is the case in the example immediately following the corollary. Thus the derivative of a nonperiodic function might be a periodic function.

It can be shown that the intrinsic derivative of any periodic function with period a can be found by differentiating any continuous function that matches its shape between $x = -\frac{1}{2}a$ and $\frac{1}{2}a$, then substituting $\operatorname{saw}_a x$ for x. The complete derivative can then be found by adding the proper delta function. For finite discontinuities we have the Second Fundamental Theorem.

Second Fundamental Theorem

Given: **1.** $p(x)$ is periodic in x with period a

2. $f(x) = p(x)$ in the interval $-\frac{1}{2}a \underset{\leftarrow}{\leq} x \underset{\rightarrow}{\leq} \frac{1}{2}a$

$$\rightarrow\rightarrow\rightarrow \quad \frac{d}{dx}p(x) = \frac{d}{du}f(u) + r\delta\left(\frac{a}{\pi}\cos\frac{\pi x}{a}\right)$$

Where: **1.** $u = \operatorname{saw}_a x$

2. $r = f(-\frac{1}{2}a + 0) - f(\frac{1}{2}a - 0)$.

Proof is given in Appendix A. If $p(x)$ is continuous at $x = \frac{1}{2}a$ (it can be discontinuous elsewhere), the theorem reduces to

$$\frac{d}{dx}p(x) = \frac{d}{du}f(u), \qquad r = 0 \tag{4.27}$$

Some examples of the use of the Second Fundamental Theorem follow.

Example 4.1. $p(x) = \text{tri } x,\ a = 2\pi,\ r = 0,\ f(u) = |u|$, giving

$$\frac{d}{dx}\text{tri } x = \frac{d}{du}|u| = \text{sgn } u$$
$$= \text{sgn}(\text{saw}_{2\pi}x) = \text{sqr } x$$

by Example 3.2.

Example 4.2. $p(x) = |\cos x|,\ a = \pi,\ r = 0,\ f(u) = \cos u$, giving

$$\frac{d}{dx}|\cos x| = \frac{d}{du}\cos u = -\sin u$$
$$= -\sin(\text{saw}_{\pi}x) = -\text{sgn}(\cos x)\sin x.$$

This is a negative pseudorectified sinewave. See Example 3.3 and Fig. 3.8. We can differentiate the result of this differentiation to obtain the second derivative of $|\cos x|$ as follows:

$$\frac{d^2}{dx^2}|\cos x| = \frac{d}{dx}[-\text{sgn}(\cos x)\sin x]$$
$$= 2\delta(\cos x)\sin^2 x - \text{sgn}(\cos x)\cos x \tag{4.28}$$
$$= -|\cos x| + 2\delta(\cos x).$$

Example 4.3. $p(x) = \text{saw}_a x,\ a = a,\ r = -a,\ f(u) = u$, giving

$$\frac{d}{dx}\text{saw}_a x = 1 - a\delta\left(\frac{a}{\pi}\cos\frac{\pi x}{a}\right).$$

From this example we also obtain the relationship

$$\frac{d}{dx}\sigma_a(x) = a\delta\left(\frac{a}{\pi}\cos\frac{\pi x}{a}\right) \tag{4.29}$$

It will be recalled that this relationship was introduced (but not proved) in Chapter 2 for the case $a = \pi$ in the form $\pi\int\delta(\cos x)dx = \sigma_\pi(x) + C$.

Example 4.4. $p(x) = |\sin x|,\ a = \pi,\ r = 0,\ f(u) = \text{sgn } u \sin u$, giving

$$\frac{d}{dx}|\sin x| = \text{sgn } u\cos u + 2\sin u\,\delta(u)$$
$$= \text{sgn}(\text{saw}_\pi x)\cos(\text{saw}_\pi x)$$
$$= \text{sgn}(\text{saw}_{2\pi} x)\cos(\text{saw}_{2\pi} x)$$
$$= \text{sqr } x \cos x.$$

The term $2 \sin u\ \delta(u) = 2\ u\ \delta(u)$ is dropped because it is a null function; $\text{sgn}(\text{saw}_\pi x) \cos(\text{saw}_\pi x)$ is a frequency-doubling expression of the type discussed in Section 3.5. Note that the period can also be taken as 2π in which case the final answer, sqr x cos x, follows more quickly. The result, sqr x cos x, is a pseudorectified cosinewave as graphed in Fig. 3.9.

Example 4.5. $p(x) = \text{sqr}\ x$, $a = 2\pi$, $r = -2$, $f(u) = \text{sgn}\ u$, giving

$$\frac{d}{dx}\text{sqr}\ x = 2\delta(u) - 2\delta(2 \cos \tfrac{1}{2}x)$$

$$= 2\delta(2 \sin \tfrac{1}{2}x) - 2\delta(2 \cos \tfrac{1}{2}x).$$

Earlier, we showed that the derivative of sqr x is $2\delta(\sin x) \cos x$. It is easy to show by graphing that these two results are the same, i.e., that $2\delta(2 \sin \tfrac{1}{2}x) - 2\delta(2 \cos \tfrac{1}{2}x) = 2\delta(\sin x)\cos x$.

Example 4.6. $p(x) = 1 - (4/\pi^2)\ \text{saw}_\pi^2 x$, $a = \pi$, $r = 0$, $f(u) = 1 - (4/\pi^2)u^2$, giving

$$\frac{d}{dx}p(x) = -\ (8/\pi^2)\ \text{saw}_\pi x.$$

$p(x)$ is diagrammed in Fig. 3.10.

The Second Fundamental Theorem is a general structured method for finding the derivatives of periodic functions. It is not necessarily the most direct method. For example, the rectified sine and cosine waves can be differentiated more directly by applying Theorem 4.1 instead.

All of the above examples deal with the general periodic function $p(x)$. The results are also adaptable to functions $p(v)$ where $v = v(x)$ by using the relationship

$$\frac{d}{dx}p(v) = \frac{d}{dv}p(v)\frac{dv}{dx} \tag{4.30}$$

Note that, although $p(v)$ is periodic in v, it is not necessarily periodic in x. An example of a function that is not periodic in x is $p(v) = |\sin v|$ with $v = x^2$. Thus it is seen how the Second Fundamental Theorem is usable with certain nonperiodic functions as well as periodic ones. In the present case, one obtains

$$\frac{d}{dx}|\sin x^2| = 2x\ \text{sqr}\ x^2 \cos x^2 \tag{4.31}$$

4.4 Applications

4.4.1 Television raster scanning

Modern television makes use of a scanning raster whose horizontal and vertical components behave in an approximate sawtooth-wave fashion.[6] In this section, we find the voltage wave necessary to produce a sawtooth current wave in a magnetic deflection yoke.

We let the desired scanning current be

$$I(t) = I_0 \mathrm{saw}_{2\pi} \omega t \tag{4.32}$$

with I_0 being a constant and ω being the angular frequency, so that the frequency is $f = \frac{1}{2}\omega/\pi$. We address the vertical deflection coils.

The equivalent circuit of vertical deflection coils is very nearly all inductance, L, and resistance, R, in series. Therefore, the expression for the required voltage is, where $\dot{I} = dI/dt$,

$$E(t) = L\dot{I}(t) + RI(t) = LI_0\omega[1 - 2\pi\,\delta(2\cos\tfrac{1}{2}\omega t)] + RI_0 \,\mathrm{saw}_{2\pi}\omega t \tag{4.33}$$

Thus the required voltage consists of the sum of a d-c component, a sawtooth wave component, and a delta function component.

This situation is impossible of physical realization because of the requirement to generate infinitely-high delta pulses of voltage; nevertheless, television is a reality. The way around this dilemma is to settle for a scanning current that is not a perfect sawtooth wave. The actual scanning current used in practice is a triangular wave of the type shown in Fig. 4.1. This function is called a *sweep function*. The portion with the more-gentle slope is the *trace*, and the portion with the steeper slope is the *retrace*. The retrace is also called the "flyback." The derivative of this waveform contains "flyback pulses" having a narrow but finite width. Pulses of this kind are discussed in Chapter 6.

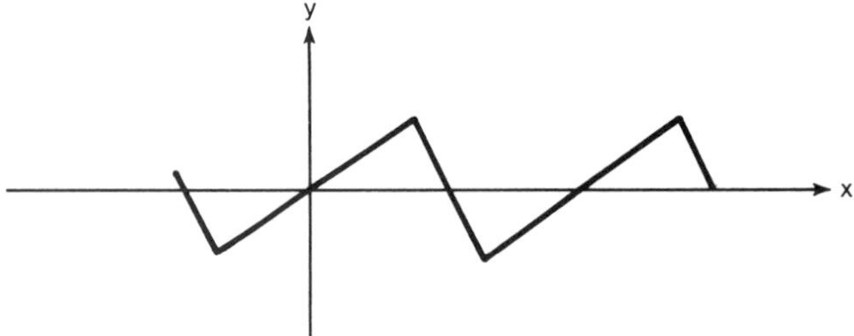

Figure 4.1 A sweep function.

[6] Also see Section 10.4.3 of Chapter 10.

4.4.2 A bouncing-ball oscillator

A perfectly elastic spherical mass m is dropped onto a horizontal, perfectly elastic immobile plate from height h at time $t = 0$. Damping is zero. Downward force is mg, assumed to be constant. Once each cycle, the mass experiences an upward impulse when it impacts the plate.

In this ideal system, the mass will oscillate indefinitely in a manner superficially resembling a full-wave rectified cosinewave—actually a parabolic wave—as shown in Fig. 3.10. The motion is, for a doubly-infinite wavetrain,

$$y = h(1 - \frac{4}{\pi^2} \operatorname{saw}_\pi{}^2 \omega t) \tag{4.34}$$

The upward force acting on the mass is

$$F = m\ddot{y} = -mg + mk\,\delta\,(\cos\omega t) \tag{4.35}$$

where $\ddot{y}$ means d^2y/dt^2.

4.4.3 A mechanical triangular wave oscillator

Figure 4.2 shows a mass-spring system with damping. The motion of the mass in this system can be characterized by the equation

$$F(t) = m\,\ddot{y} + c\,\dot{y} + k\,y \tag{4.36}$$

where $F(t)$ is the driving force, m is the mass, c is the damping factor, k is the spring constant, and y is the displacement from rest position.

If we want the motion of the mass to be such that

$$y = A(\operatorname{tri} \omega t - \tfrac{1}{2}\pi) \tag{4.37}$$

then it is a simple matter to show that the driving force must be

$$F(t) = 2\text{m}A\omega\cos\omega t\;\delta(\sin\omega t) + cA\omega \operatorname{sqr} \omega t + ky \tag{4.38}$$

Just as was the case with the bouncing-ball oscillator, the delta pulse train term (the first term) of Eq. (4.38) is the impulsive-force train produced when the mass—if it is perfectly elastic—impacts fixed limit stops that are also perfectly elastic. Also, the *impulse* produced upon each impact

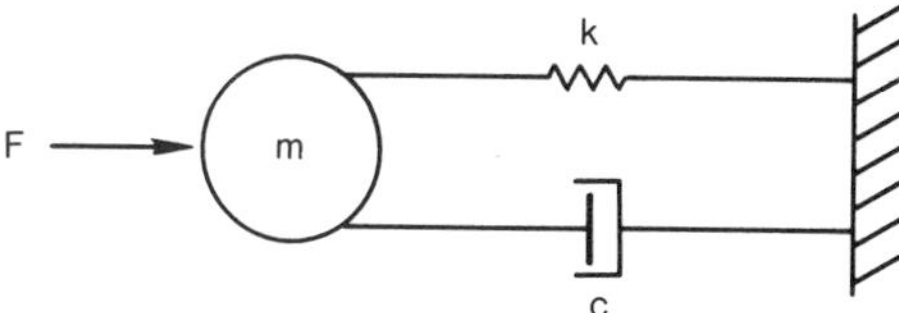

Figure 4.2 A mass-spring system with damping.

(of the mass with a limit stop) is equal to the *content* of a given delta pulse, relative to t.[7]

The damper-dependent term, $cA\omega$ sqr ωt, is constant in magnitude because the velocity of the mass is constant in magnitude, i.e., its speed is constant. Finally, the spring-dependent term, ky, duplicates the desired motion.

In such an ideal system, if we let $c = k = 0$, then the mass—once set in motion—will continue to oscillate indefinitely in the mode described by Eq. (4.37). In a real physical system there are inevitable losses; but if the lost energy is continually replaced, an ideal system can be closely simulated.

The constant A is given by

$$A = B/\pi \tag{4.39}$$

where B is the distance between limit stops for a point mass. For a nonpoint mass, B is the distance between the limit stops minus the diameter of the mass. B is the *double amplitude*. The parameter ω, angular frequency, is given by

$$\omega = |\dot{y}|/A = |\dot{y}|\pi/B \tag{4.40}$$

where $|\dot{y}|$ is the constant speed of the mass.

For a sinusoidal oscillator with $c = 0$, there is a particular frequency at which a mass-spring system oscillates naturally. This is its resonant frequency. In the case of the triangular wave oscillator with $c = k = 0$, the system will oscillate "naturally" at any frequency. The frequency depends on the initial speed given to the mass, according to Eq. (4.40). In this case "resonance" is a meaningless concept. In this respect the triangular wave oscillator with $c = k = 0$ is an analog of the flywheel.

4.4.4 A hybrid oscillator

A sinusoidal system with limit stops is of interest; let us examine the case where the motion of a mass is described by the symmetrically-interrupted cosinewave of Fig. 3.13. The equation of motion of the mass is

$$y = A \cos \frac{\pi(u + \alpha)}{\pi + 2\alpha}, \quad u = \text{tri}\, \omega t \tag{4.41}$$

When $\alpha = 0$, Eq. (4.41) describes a sinusoidal oscillator. If elastic limit stops are inserted into the system symmetrically so that $\alpha > 0$, then

[7] From elementary physics it is recalled that *impulse*, $\int F(t)dt$, is equal to the *change in momentum*, $\Delta(mv)$.

the oscillations of the mass are interrupted and we have a *hybrid oscillator*, one with some properties of a sinusoidal oscillator and some of a triangular wave oscillator.

From Eq. (4.41) we have

$$\dot{y}/A = -\frac{\omega\pi}{\pi + 2\alpha} \sin \frac{\pi(u + \alpha)}{\pi + 2\alpha} \operatorname{sqr} \omega t \tag{4.42}$$

$$\ddot{y}/A = -\frac{(\omega\pi)^2}{(\pi + 2\alpha)^2} \cos \frac{\pi(u + \alpha)}{\pi + 2\alpha} - \frac{\omega^2\pi}{\pi + 2\alpha} \cos \omega t \, \delta(\sin\omega t) \sin \frac{\pi(u + \alpha)}{\pi + 2\alpha} \tag{4.43}$$

From Eqs. (4.41) through (4.43) we can then obtain the desired force equation by use of Eq. (4.36). Resonance for $c = 0$ occurs when

$$k = m \frac{(\omega\pi)^3}{(\pi + 2\alpha)^2} \tag{4.44}$$

so that the angular frequency at resonance is

$$\rightarrow\rightarrow\rightarrow \quad \omega_r = (1 + 2\alpha/\pi)\omega_{rs} \tag{4.45}$$

where ω_{rs} is the resonant angular frequency of the sinusoidal oscillator with $\alpha = 0$, namely,

$$\omega_{rs} = (k/m)^{1/2} \tag{4.46}$$

Equation (4.45) shows that the resonant frequency of the hybrid oscillator ($\alpha > 0$) is greater than that of a sinusoidal one with the same mass and spring constant. This means that such an oscillator is capable of storing a greater amount of energy for a given amplitude of oscillation.

If we let the double amplitude be B as before, then

$$\alpha = \operatorname{Cos}^{-1} \frac{B}{2A} \tag{4.47}$$

and we have

$$\omega_r = \left(1 + \frac{2}{\pi} \operatorname{Cos}^{-1} \frac{B}{2A}\right) \omega_{rs} \tag{4.48}$$

If we let the amplitude be constant as a is varied, then A must vary as $\sec \alpha$, or

$$A = \tfrac{1}{2} B \sec \alpha \tag{4.49}$$

from Eq. (4.47).

Example. The resonant angular frequency of the hybrid semi-interrupted oscillator described by the equation of Example 3.9 is

$$\omega_r = (1 + \alpha/\pi)\, \omega_{rs}$$

Example. The resonant angular frequency of the hybrid asymmetrically-interrupted oscillator with interruptions α and β is

$$\omega_r = (1 + \alpha/\pi + \beta/\pi)\omega_{rs}$$

4.5 Super Resonance

4.5.1 General

The mass-spring system operating as a hybrid oscillator has been shown to have a resonant frequency greater than that of the same system operating as a sinusoidal oscillator. This condition is called *super resonance*. It is a condition produced when a linear oscillatory system is operated in a nonlinear mode. The nonlinearity is caused by the immobile, elastic limit stops in the case just discussed.

4.5.2 A mechanical binary system

This kind of nonlinearity can also be produced by bringing two sinusoidal oscillators together so that they mutually interfere.

Consider, for example, the "binary" torsion pendulum system of Fig. 4.3. If these two pendulums are set in motion with double amplitudes less than 180°, and are 180° out of phase with each other, they will operate as two independent sinusoidal oscillators. But if we attempt to increase their peak-to-peak amplitudes (equally) to more than 180°, then they will operate as two hybrid oscillators because their bobs will impact each other twice each cycle, along the dotted line in Fig. 4.3. Their frequencies will be increased, signalling a condition of super resonance.

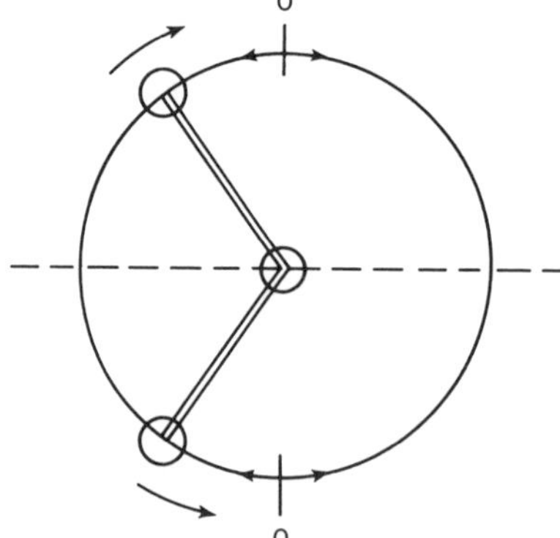

Figure 4.3 A binary torsion pendulum system.

This happens even though the two physical pendulums have not been changed; they are still described by the same linear springs, constant masses and constant moments of inertia as before. The super resonance condition is *solely a result of the new mode of excitation*. Energy is exchanged between the two bobs when they impact, causing the nonlinear action and the super resonance condition.

4.5.3 An electrical binary system

Super resonance can also be produced in linear electrical resonant circuits if we arrange to have energy periodically exchanged between two such circuits. Fig. 4.4 shows one way we might do this.

If the two circuits of Fig. 4.4 are excited in synchronism but 180° out of phase, then they will operate as two separate sinusoidal oscillators as long as the amplitudes (assumed equal) are too small to cause the zener diodes to conduct. But if the amplitudes are increased (equally) so that the zener diodes begin to conduct on peaks, then energy will be exchanged between the two LC circuits, and a super resonant mode will begin.

That is, these two tuned circuits, even though they have constant values of L and C, will act as if they had been tuned to a higher frequency *simply as a result of the new mode of excitation*. This circuit is the electrical analog of the mechanical system described in Subsection 4.5.2. The currents flowing in the two LC circuits will have the form of symmetrically interrupted sinusoids.

We take this path of analogies one step further in the following subsection.

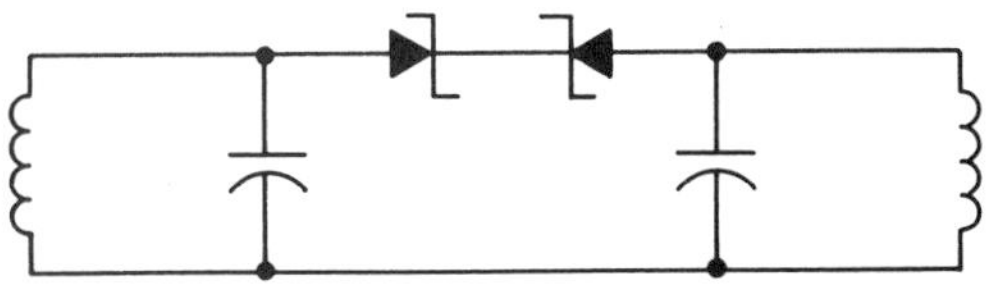

Figure 4.4 An electrical binary system.

4.5.4 The "resonant speed" of space

The constancy of the speed of light in free space leads to the now-classic equation of special relativity, $\gamma = (1 - \beta^2)^{-1/2}$, which describes relativistic effects, with β being "normalized velocity." Among these effects are "space contraction," "time dilation," and "mass increase with velocity." That equation for γ can be "explained" by likening the speed of light in free space to a resonance condition.

An electronic circuit analog was announced by Schelkunoff in 1938[8] in which it was shown that the intrinsic impedance of free space is analogous to the characteristic impedance of a network, and the Q of a lossy medium is analogous to the Q of a network. That analog is in common use today in antenna design.

That electronic circuit analog has been developed further by the author in accordance with the ideas that the permittivity and permeability of free space are analogous to the capacitance and inductance, respectively, of a tuned circuit. In this extension of the analogy, the speed of light becomes analogous to the resonant frequency of a tuned circuit.[9]

While apparently it is not possible to decrease the value of either the permittivity or permeability of free space and thereby "tune" it to a higher "resonant speed", the following question nevertheless suggests itself: Can a region of free space be *effectively* tuned to a higher resonant speed by exciting it in the manner of the binary systems described above? This question remains unanswered.

4.6 Extrinsic Derivatives

Previous sections of this chapter deal with derivatives of mathematically ideal functions. For example, the derivative of $|x|$ has been shown to have a finite discontinuity, with infinitesimal hysteresis, at $x = 0$.

It is impossible to generate such ideal functions, as is pointed out in the example of Subsection 4.4.1. In the physical applications, "broken functions" have some rounding[10] at the breaks; "finitely-discontinuous" functions are not really discontinuous; delta pulses are replaced by pulses having a finite width and amplitude; and the need for infinitesimal hysteresis vanishes.

The derivatives of such nonideal functions are said to be *extrinsic derivatives*, to distinguish them from the derivatives of ideal functions. This is a convenient nomenclature only, and does not indicate a fundamentally different kind of derivative. It simply indicates that we have

[8] S. A. Schelkunoff, "The Impedance Concept and its Application to Problems of Reflection, Refraction, Shielding, and Power Absorption," *The Bell System Technical Journal,* XVII, no. 1 (January 1938), pp. 17–48.

[9] H. B. Tilton, "An Electronic Analog of Relativistic Space," *Electron and Ion Beam Science and Technology* (New York: The Electrochemical Society, Inc., 1968), pp. 3–14.

[10] The amount of rounding depends on the high-frequency response or slew rate of the circuits involved.

taken the derivative of a nonideal function. Thus the extrinsic derivative of $|x|$ is a continuous "ramp" function of the type discussed in Chapter 6. This derivative is the same as the intrinsic derivative of a nonideal rounded $|x|$ function.

General methods of dealing with such nonideal functions are developed in Chapter 6.

5

HOW TO INTEGRATE RELAXATION FUNCTIONS

5.1 Integration as the Inverse of Differentiation; the Definite Integral as an Area

In the differential calculus, differentiation is taken as basic; integration is then defined as the inverse operation. In the integral calculus the situation is reversed: integration is taken as basic and differentiation is the inverse operation. In either case, differentiation senses the slope of a curve. The situation for differentiation is the same in relaxation analysis except that the concept of slope is extended to encompass discontinuities as is shown in the previous chapter. But what is integration?

There are two kinds of integration, as we know from elementary calculus: indefinite integration and definite integration. The definite integral is the area under the curve between the limits of integration. The situation for definite integration is the same in relaxation analysis. An indefinite integral can be interpreted as an *incremental area* or simply as *the function whose derivative is the given function*, i.e., as the inverse of differentiation. With this second interpretation, the concept of *integral at a point* is viable in the same way that *derivative at a point* (the slope of the tangent line at the point) is a viable concept of derivative.

The concept of *integral* in relaxation analysis is the same as in elementary calculus. There is less difficulty here than was the case for differentiation. Definite integrals of finitely discontinuous functions appear throughout standard analysis (for example, in finding Laplace transforms).

To these we add indefinite integrals and integrals of delta functions. These latter integrals were discussed in Chapter 2.

Since the derivative of a continuous-broken function is a discontinuous function, it follows that the integral of a discontinuous function is a continuous-broken function. It is also easy to show that the integral of a continuous-broken function is a continuous-smooth function, and that the integral of a continuous-smooth function is also continuous-smooth.

This chapter presents methods of finding indefinite integrals or *integral functions* of relaxation functions. It also presents rules governing definite integration.

5.2 How to Integrate Nonperiodic Functions

In standard analysis, integration of a continuous-broken or discontinuous nonperiodic function is normally performed in a piecewise manner. For example, the integral function of $|x|$ might be written in piecewise form as follows:

$$\begin{aligned} \int |x| dx &= \int x\, dx = \tfrac{1}{2}x^2 + C \qquad \text{when } x > 0 \\ \int |x|\, dx &= \int (-x)\, dx = -\tfrac{1}{2}x^2 + C \qquad \text{when } x < 0 \end{aligned} \tag{5.1}$$

The astute student might then recognize both results as conforming to the single expression $\frac{1}{2}|x|x + C$.

It is a goal of relaxation analysis to obtain the integrals of expressions such as $|x|$ directly, as single "compact" expressions. Toward this end we find that standard integral tables are useful. Some of these integrals are given in Table 5.1. Other integrals that cannot be found in standard integral tables are developed throughout this chapter. These are collected together in Appendix I.

As an example of the use of standard integrals for integrating relaxation functions, we examine the function $|x|$. Integrating by parts, letting $u = |x|$ and $dv = dx$, we obtain

$$\int |x| dx = |x|x - \int x \operatorname{sgn} x\, dx = |x|x - \int |x| dx \tag{5.2}$$

Equation (5.2) is readily solved to give the final result

$$\int |x| dx = \tfrac{1}{2}|x|x + C \tag{5.3}$$

Checking the result by differentiating gives

$$\frac{d}{dx}(\tfrac{1}{2}|x|x + C) = \tfrac{1}{2}|x| + \tfrac{1}{2}x\frac{d}{dx}|x| = |x| \tag{5.4}$$

as required.

TABLE 5.1 Integrals

With the aid of the relationships $\frac{du}{dx}dx = \frac{dx}{dx}du = du$ and $\int du = u$, the following integrals are obtained by applying the integration operation $\int()dx$ to the derivatives of Table 4.1.

GENERAL FORMS

1. From No. 1 of Table 4.1 (integral of a sum)

$$\int(du + dv) = u + v$$

or

$$\int(f + g)\,dx = \int f dx + \int g dx$$

2. From No. 2 of Table 4.1 (integration by parts)

$$\int u dv = uv - \int v du$$

3. From No. 3 of Table 4.1

$$\int \frac{u}{v^2}dv = -u/v + \int \frac{1}{v}du, \qquad v \neq 0$$

4. From No. 4 of Table 4.1

$$\int \frac{u^2}{(u+v)^2}dv = \frac{uv}{u+v} - \int \frac{v^2}{(u+v)^2}du, \qquad u + v \neq 0$$

SPECIFIC FORMS

5. From Nos. 7 and 8 of Table 4.1

$$\int 0\,dx = \text{constant (real, imaginary, or complex)}$$

Note: Because of #5, an arbitrary constant should be added to each specific indefinite integral.

6. From No. 9 of Table 4.1

$$\int x\,dx = \tfrac{1}{2}x^2$$

7. From No. 10 of Table 4.1

$$\int \cos x\,dx = \sin x$$

8. From No. 11 of Table 4.1

$$\int \sin x\,dx = -\cos x$$

9. From No. 12 of Table 4.1

$$\int e^x dx = e^x$$

Another way to integrate $|x|$ is to use the form $\int(X)^{1/2}dx$ where $X = a + bx + cx^2$ with $a = b = 0$ and $c = 1$. This form appears in tables of integrals. It leads to the same result as before.

Next we consider $\int$ sgn $x\,dx$. Let u = sgn x and $dv = dx$, giving

$$\int \operatorname{sgn} x\,dx = x \operatorname{sgn} x - 2\int x\delta(x)dx = |x| + C \qquad (5.5)$$

The right-hand integral is a constant because $x\delta(x)$ is a null function.

Another way to integrate the signum function is as follows: let $u = \frac{1}{2}x^2$ and write

$$\begin{aligned}\int \operatorname{sgn} x\, dx &= \int \frac{x dx}{(x^2)^{1/2}} = \int \frac{du}{(2u)^{1/2}} = (2u)^{1/2} + C \\ &= (x^2)^{1/2} + C = |x| + C\end{aligned} \tag{5.6}$$

Still another way is to write sgn $x\, dx = d|x|$ from Eq. (4.2), giving

$$\int \operatorname{sgn} x\, dx = \int d|x| = |x| + C \tag{5.7}$$

If the integral of $f(x)$ exists, then the integral of $f(x)$ sgn x exists. Write

$$y = \int f(x) \operatorname{sgn} x\, dx \tag{5.8}$$

This becomes

$$y = F(x) \operatorname{sgn} x - 2\int F(x)\delta(x)dx \tag{5.9}$$

where $F(x) = \int f(x)dx$. Finally,

$$\int f(x) \operatorname{sgn} x\, dx = \operatorname{sgn} x \int_0^x f(x)dx + C \tag{5.10}$$

If the integral of $f(x)$ exists, then the integral of $f(x)$ step$(x - a)$ exists where step $(x - a)$ is any step function with step at $x = a$. It is

$$\rightarrow\rightarrow\rightarrow \quad \int f(x) \operatorname{step}(x - a)\, dx = \operatorname{step}(x - a)\int_a^x f(x)dx + C \tag{5.11}$$

Proof is similar to that for $\int f(x)$ sgn $x\, dx$.

As a check of Eq. (5.11) we consider this example. A constant can be considered to be a step function with step of zero rise. From Eq. (5.11) we have

$$\begin{aligned}\int f(x) \text{ const } dx &= \text{const} \int_a^x f(x)dx + C \\ &= \text{const } [F(x) - F(a)] + C \\ &= \text{const } F(x) - \text{const } F(a) + C \\ &= \text{const} \int f(x)dx + C'\end{aligned} \tag{5.12}$$

Examples of the use of Eq. (5.10) are

$$\begin{aligned}\int |x|dx &= \int x \operatorname{sgn} x\, dx = \operatorname{sgn} x \int_0^x x\, dx + C \\ &= \operatorname{sgn} x\, \tfrac{1}{2}x^2 + C = \tfrac{1}{2}|x|x + C\end{aligned} \tag{5.13}$$

$$\int \sin x \operatorname{sgn} x\, dx = (1 - \cos x) \operatorname{sgn} x + C \tag{5.14}$$

$$\int x^2 \operatorname{sgn} x\, dx = (x^3/3)\operatorname{sgn} x + C \tag{5.15}$$

5.3 How to Integrate Nonsinusoidal Periodic Functions

The central theme of this section is a theorem that permits the integration of a wide range of periodic functions. However, before we present that theorem, some examples are given of the use of classical integration techniques with specific periodic relaxation functions.

We begin with a particularly simple integration. To find the integral of $|\sin x| \cos x$, let $u = \sin x$.

$$\int |\sin x| \cos x\, dx = \int |u| du = \tfrac{1}{2}|u|u + C \tag{5.16}$$
$$= \tfrac{1}{2}|\sin x| \sin x + C$$

The sawtooth wave $\operatorname{saw}_\pi x$ can be integrated using a formula from elementary calculus. We write

$$\int \operatorname{saw}_\pi x\, dx = \int \operatorname{Tan}^{-1}\tan x\, dx = \int \operatorname{Tan}^{-1} u \frac{du}{1 + u^2} \tag{5.17}$$

where $u = \tan x$. From standard tables of integrals, this integrates to $\frac{1}{2}(\operatorname{Tan}^{-1}u)^2$, giving finally

$$\int \operatorname{saw}_\pi x\, dx = \tfrac{1}{2} \operatorname{saw}_\pi^2 x + C \tag{5.18}$$

Sine-derived and cosine-derived squarewaves can also be integrated using classical methods. For the sine-derived squarewave we have

$$\int \operatorname{sqr} x\, dx = \int \frac{\sin x\, dx}{(1 - \cos^2 x)^{1/2}} = \int \frac{du}{(1 - u^2)^{1/2}} \tag{5.19}$$

where $u = -\cos x$. This integrates to $-\operatorname{Cos}^{-1}u$ (or $\operatorname{Sin}^{-1}u$), giving finally

$$\int \operatorname{sqr} x\, dx = -\operatorname{Cos}^{-1}(-\cos x) + C = \operatorname{Cos}^{-1}\cos x + C = \operatorname{tri} x + C \tag{5.20}$$

Similarly, for the cosine-derived squarewave, we let $u = \sin x$, giving

$$\int \operatorname{sgn}(\cos x)dx = \operatorname{Sin}^{-1}\sin x + C' = \operatorname{tri}(x + \tfrac{1}{2}\pi) + C \tag{5.21}$$

The same result can be obtained by integrating $\operatorname{sgn}[\sin(x + \frac{1}{2}\pi)]$, using the integral for $\operatorname{sqr} u$.

We now integrate the full-wave rectified cosine and sine waves. We write $\int|\cos x|dx = \int \mathrm{sgn}(\cos x)\cos x\, dx$ and integrate by parts, letting $u = \mathrm{sgn}(\cos x)$ and $dv = \cos x\, dx$, to give

$$\int|\cos x|dx = \mathrm{sgn}(\cos x)\sin x + \int 2\delta(\cos x)dx \qquad (5.22)$$

The latter integral is the cosine-derived regular staircase function $\frac{1}{\pi}\sigma_\pi(x)$ times 2, so that finally

$$\int|\cos x|dx = \mathrm{sgn}(\cos x)\sin x + (2/\pi)\sigma_\pi(x) + C \qquad (5.23)$$

Similarly,

$$\int|\sin x|dx = \int \mathrm{sqr}\, x \sin x\, dx = -\cos x\, \mathrm{sqr}\, x + \int 2\delta(\sin x)dx \qquad (5.24)$$

where the latter integral is the sine-derived regular staircase function $\frac{1}{\pi}\sigma_\pi(x - \frac{1}{2}\pi)$ times 2, so that

$$\int|\sin x|dx = -\mathrm{sqr}\, x \cos x + (2/\pi)\, \sigma_\pi(x - \tfrac{1}{2}\pi) + C \qquad (5.25)$$

In the above two examples $\mathrm{sgn}(\cos x)\sin x$ and $\mathrm{sqr}\, x \cos x$ are the pseudorectified sine and cosine waves, respectively.

The above integrations, while not overly difficult, rely on methods that are tailored to the specific problem at hand. We now present a theorem that is a structured method generally applicable to the integration of periodic functions. It facilitates the integration of functions that are integrable only with difficulty—if at all—by classical techniques. Since it is a structured method, it is particularly well-suited to computer implementation, as is the Second Fundamental Theorem. We begin with a useful lemma.

Lemma 5.1. If the average value of a periodic function with any period a is zero, then its integral function is also periodic with period a. Proof is given in Appendix A. That integral function can be found by first integrating any function that matches the given function between $x = -\frac{1}{2}a$ and $\frac{1}{2}a$, then substituting $\mathrm{saw}_a x$ for x. This concept is formalized and generalized in the Third Fundamental Theorem.

Third Fundamental Theorem

Given: **1.** $p(x)$ is periodic in x with period a
2. $f(x) = p(x)$ in the interval $-\frac{1}{2}a \underset{\leftarrow}{\leq} x \underset{\rightarrow}{\leq} \frac{1}{2}a$

$\rightarrow\rightarrow\rightarrow \quad \int p(x)dx = \int f(u)du + \bar{p}\sigma_a(x)$

Where: **1.** $u = \text{saw}_a x$
2. $\sigma_a(x) = x - u$
3. $\overline{p}$ is the average value of $p(x)$.

Proof of the theorem is given in Appendix A.

The average value of a periodic function is also called the *d-c component*. If the average value is zero, the theorem reduces to

$$\int p(x)dx = \int f(u)du, \qquad \overline{p} = 0 \tag{5.26}$$

Some examples of the use of the Third Fundamental Theorem follow.

Example 5.1. $p(x) = \text{saw}_a x$, $a = a$, $\overline{p} = 0$, $f(u) = u$, giving

$$\int \text{saw}_a x \, dx = \int u \, du = \tfrac{1}{2}u^2 + C$$
$$= \tfrac{1}{2}\text{saw}_a^2 x + C$$

Example 5.2. $p(x) = \text{sqr } x$, $a = 2\pi$, $\overline{p} = 0$, $f(u) = \text{sgn } u$, giving

$$\int \text{sqr } x \, dx = \int \text{sgn } u \, du = |u| + C$$
$$= |\text{saw}_{2\pi} x| + C$$
$$= \text{tri } x + C, \text{ by Example 3.1}$$

Example 5.3. $p(x) = \text{tri } x$, $a = 2\pi$, $\overline{p} = \tfrac{1}{2}\pi$, $f(u) = |u|$, giving

$$\int \text{tri } x \, dx = \int |u| du + \tfrac{1}{2}\pi \, \sigma_{2\pi}(x) = \tfrac{1}{2} \text{tri } x \, \text{saw}_{2\pi} x + \tfrac{1}{2}\pi\sigma_{2\pi}(x) + C$$

This is a nonperiodic function.

Example 5.4. $p(x) = |\cos x|$, $a = \pi$, $\overline{p} = 2/\pi$, $f(u) = \cos u$, giving

$$\int |\cos x| \, dx = \int \cos u \, du + (2/\pi) \, \sigma_\pi(x)$$
$$= \sin(\text{saw}_\pi x) + (2/\pi) \, \sigma_\pi(x) + C$$
$$= \text{sgn}(\cos x) \sin x + (2/\pi) \, \sigma_\pi(x) + C$$

by Example 3.3.

The result is a nonperiodic function. The function $\text{sgn}(\cos x) \sin x + (2/\pi) \, \sigma_\pi(x)$ is graphed in Fig. 5.1. The block diagram of a staircase generator based on this method is shown in Fig. 5.2.

If the d-c component is subtracted from $|\cos x|$, the resulting *a-c*

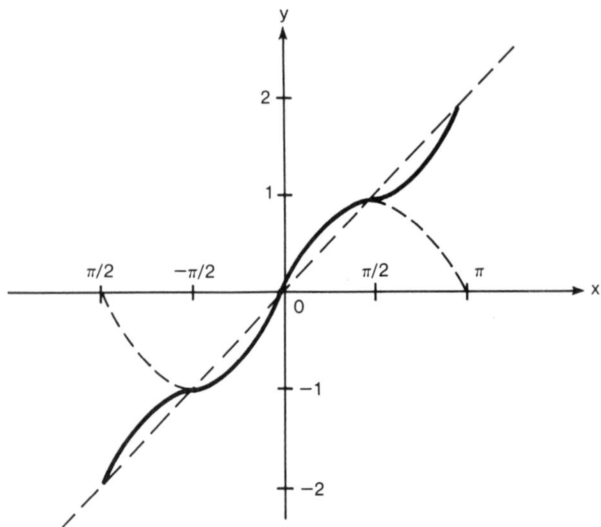

Figure 5.1 The integral function of a full-wave rectified cosinewave.

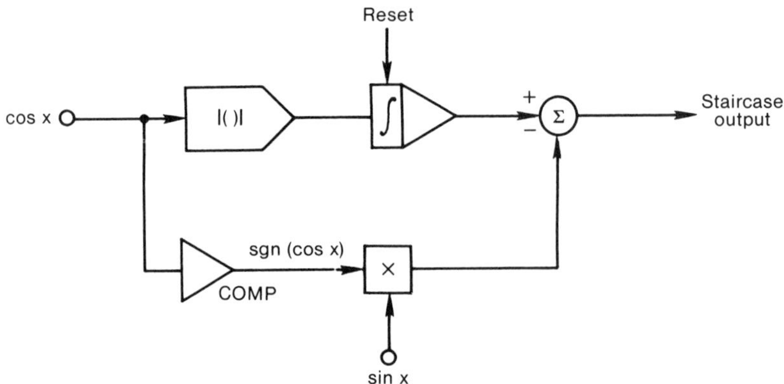

Figure 5.2 Block diagram of staircase function generator based on Example 5.4.

component—when integrated—produces a periodic function by Lemma 5.1. This is shown in the following example.

Example 5.5. $p(x) = |\cos x| - 2/\pi,\ a = \pi,\ \bar{p} = 0,$
$f(u) = \cos u - 2/\pi$, giving
$$\int p(x)\,dx = \sin(\text{saw}_\pi x) - (2/\pi)\text{saw}_\pi x + C.$$

This is the skewed sinuous wave shown in Fig. 5.3 for $C = 0$. Its amplitude is $\left(1 - \frac{4}{\pi^2}\right)^{1/2} - \frac{2}{\pi}\text{Cos}^{-1}\frac{2}{\pi}$ (about 0.21) occurring at $x = \cos^{-1}\frac{2}{\pi}$ (about 50 degrees for the "first" peak). This example shows one way to operate

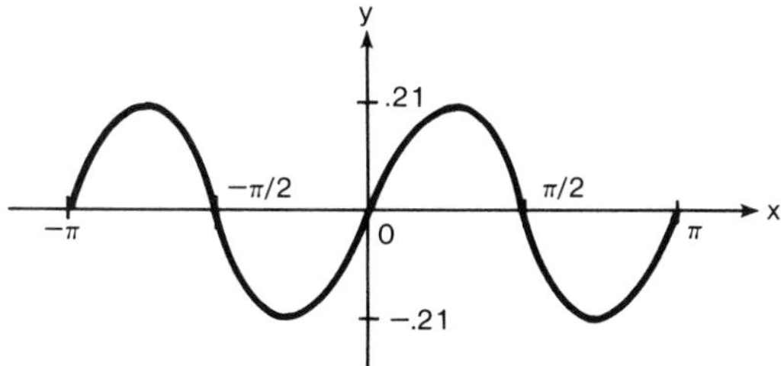

Figure 5.3 A skewed sinuous wave. The integral function of the a-c component of a full-wave rectified cosinewave.

on a sinusoid to obtain an approximate sinusoid of twice the frequency. The block diagram of a frequency doubler based on this property is shown in Fig. 5.4. In this system a sinusoidal input gives a sinuous output of twice the frequency. An advantage of this method of frequency doubling over the more common method of squaring is that a multiplier is not required.

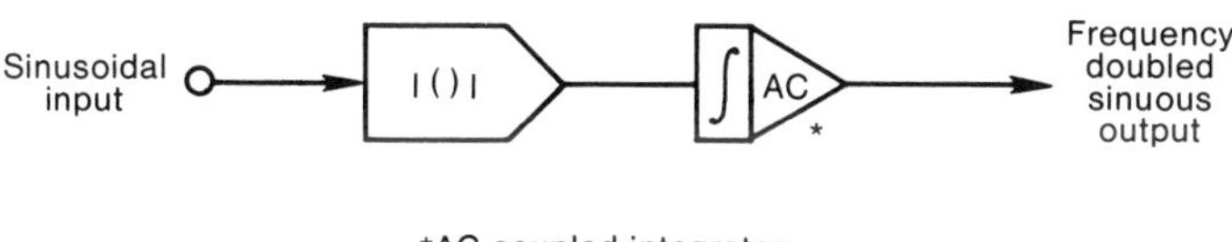

*AC coupled integrator

Figure 5.4 Block diagram of sinusoid-to-sinuous frequency doubler based on Example 5.5.

We now find the integral function of the full-wave rectified sinewave by this method.

Example 5.6. $p(x) = |\sin x|$, $a = \pi$, $\bar{p} = 2/\pi$, $f(u) = \operatorname{sgn} u \sin u$, giving

$$\int |\sin x|\, dx = \int \operatorname{sgn} u \sin u \, du + (2/\pi)\,\sigma_\pi(x)$$

$$= \operatorname{sgn} u\,(1 - \cos u) + (2/\pi)\,\sigma_\pi(\mathrm{x}) + C$$

$$= -\operatorname{sgn} u \cos u + \operatorname{sgn} u + (2/\pi)\,\sigma_\pi(x) + C$$

$$= -\operatorname{sqr} x \cos x + [\operatorname{sqr} 2x + (2/\pi)\,\sigma_\pi(x)] + C$$

because

$$\operatorname{sgn}(\operatorname{saw}_\pi x)\cos(\operatorname{saw}_\pi x)$$

$$= \operatorname{sgn}(\operatorname{saw}_{2\pi} x)\cos(\operatorname{saw}_{2\pi} x)$$

$$= \operatorname{sqr} x \cos x$$

by the frequency-doubling relationship.

The staircase function in brackets, sqr $2x + (2/\pi)\sigma_\pi(x)$, equals the staircase function of Eq. (5.25), $(2/\pi)\sigma_\pi(x - \frac{1}{2}\pi)$, except for an additive constant. This can be shown graphically. It can also be shown analytically by use of the Fourier series expansions

$$\text{Ctn}^{-1}\text{ctn}\, x = \frac{\pi}{2} + \text{saw}_\pi(x - \tfrac{1}{2}\pi) \tag{5.27}$$

$$= \sum_{n=1}^{\infty} \frac{1}{n} \sin 2nx \tag{5.28}$$

$$\text{Tan}^{-1}\text{tan}\, x = \text{saw}_\pi x = \sum_{n=1}^{\infty} \frac{(-1)^{n+1}}{n} \sin 2nx \tag{5.29}$$

$$\text{sqr}\, 2x = \frac{4}{\pi} \sum_{n=1,3,\ldots}^{\infty} (1/n) \sin 2nx$$

Example 5.7. This example is the same as Example 5.6 except we let $a = 2\pi$ so that $u = \text{saw}_{2\pi} x$ and $\sigma_a(x) = \sigma_{2\pi}(x)$. This gives

$$\begin{aligned}\int |\sin x|\, dx &= \text{sgn}\, u\,(1 - \cos u) + (2/\pi)\,\sigma_{2\pi}(x) + C \\ &= \text{sqr}\, x\,(1 - \cos x) + (2/\pi)\,\sigma_{2\pi}(x) + C \\ &= -\,\text{sqr}\, x \cos x + \text{sqr}\, x + (2/\pi)\,\sigma_{2\pi}(x) + C \\ &= -\,\text{sqr}\, x \cos x + \text{sqr}\, 2x + (2/\pi)\,\sigma_\pi(x) + C\end{aligned}$$

by Eq. (3.38).

Another way to integrate $|\sin x|$ is to recognize that $\sin x = \cos(x - \frac{1}{2}\pi)$ and use the result of Example 5.4. Thus

$$\int |\sin x| dx = \int |\cos v| dv \tag{5.30}$$

where $v = x - \frac{1}{2}\pi$. By this means we immediately obtain

$$\int |\sin x| dx = \text{sgn}(\cos v)\sin v + (2/\pi)\sigma_\pi(v) + C \tag{5.31}$$

with $\text{sgn}(\cos v) = \text{sgn}(\sin x)(-\cos x)$, so that finally

$$\rightarrow\rightarrow\rightarrow \quad \int |\sin x| dx = -\text{sqr}\, x \cos x + (2/\pi)\sigma_\pi(x - \tfrac{1}{2}\pi) + C \tag{5.32}$$

5.4 How to Perform Definite Integration of Relaxation Functions

5.4.1 Integration of intrinsic vs. complete derivatives

The concepts of "intrinsic derivative" and "complete derivative" are important to definite integration (as well as to indefinite integration). The two kinds of derivative have contrasting properties in the following sense.

Let $F(x)$ be a finitely discontinuous function. Let its derivative be

$$\frac{d}{dx}F(x) = f(x) = \phi(x) + \Delta \tag{5.33}$$

where $f(x)$ is the complete derivative, $\phi(x)$ is the intrinsic derivative, and Δ represents the requisite delta function.

It is a property of the intrinsic derivative, $\phi(x)$, that

$$\int_a^b \phi(x)dx = \mathrm{F}(x)\Big|_a^b \qquad ((5.34))^1$$

holds only if the interval ab includes no discontinuities in $F(x)$. And it is a property of the complete derivative, $f(x)$, that

$$\int_a^b f(x)dx = F(x)\Big|_a^b \tag{5.35}$$

holds over any interval ab.

For example, let $F(x) = \mathrm{saw}_\pi x$, $\phi(x) = 1$, $f(x) = 1 - \pi\delta(\cos x)$. We have, integrating $\phi(x)$,

$$\int_a^b \phi(x)dx = b - a \tag{5.36}$$

which equals $\mathrm{saw}_\pi x\Big|_a^b$ only when the interval ab does not contain a discontinuity in $\mathrm{saw}_\pi x$; because then and only then does

$$\mathrm{saw}_\pi x\Big|_a^b = b - a \qquad ((5.37))$$

By contrast, from Eq. (5.35) we have

$$\int_a^b [1 - \pi\delta(\cos x)]\mathrm{dx} = [\mathrm{x} - \sigma_\pi(\mathrm{x})]_a^b = \mathrm{saw}_\pi \mathrm{x}\Big|_a^b \tag{5.38}$$

which holds over any inverval ab.

In the above equations the notation $F(x)]_a^b = F(x)|_a^b$ has its usual meaning

$$F(x)\Big|_a^b = F(x)\Big|_{x=a}^{x=b}. \tag{5.39}$$

The precise meaning of this notation is made clear in the sections immediately following. We sometimes use a square bracket instead of the customary vertical bar.

[1] Double parentheses are used to flag equations having conditional validity.

5.4.2 Definite integration across discontinuities and breaks

Consider the generalized discontinuous function depicted in Fig. 5.5. If we wish to integrate $f(x)$ between a and b, then in standard analysis we might integrate $f_1(x)$ from a to c, then integrate $f_2(x)$ from b to c, and subtract. However, if we have a single expression for the entire discontinuous function (which is what relaxation analysis is all about!), then it is more convenient to integrate $f(x)$ directly from a to b or from b to a.

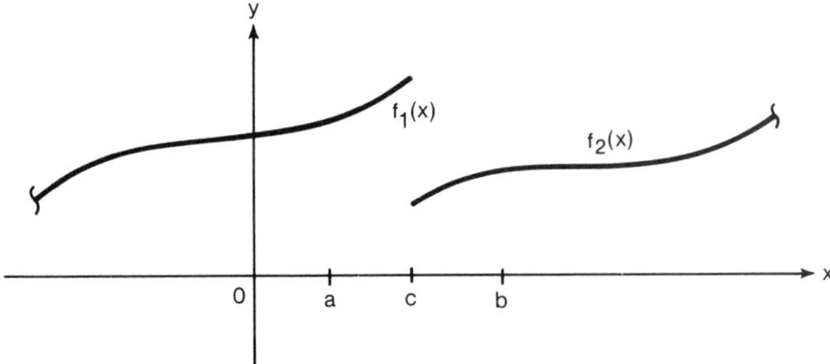

Figure 5.5 A generalized discontinuous function with a single discontinuity.

We define the definite integral as in Eq. (5.35) with

$$F(x)\Big|_a^b = F(b) - F(a) \tag{5.40}$$

Equations (5.35) and (5.40) also apply to broken functions as well as finitely-discontinuous ones, for these can be considered special cases of discontinuous functions wherein the magnitude of the discontinuity is zero. Equations (5.35) and (5.40) apply equally as well to integration across multiple discontinuities and breaks. Two examples follow

Example 5.8. Integration across a break:

$$\int_{-2}^{2} |x|dx = \tfrac{1}{2}|x|x\Big]_{-2}^{2} = 2 - (-2) = 4$$

Example 5.9. Integration across two discontinuities:

$$\int_{-\pi/2}^{7\pi/4} \operatorname{sqr} x \, dx = \operatorname{tri} x\Big|_{-\pi/2}^{7\pi/4} = \tfrac{1}{4}\pi - \tfrac{1}{2}\pi = -\tfrac{1}{4}\pi$$

Graphical checks of the above two results show that they give the areas under the respective curves.

5.4.3 Definite integration between discontinuities

To complete our investigation of the definite integration of discontinuous and broken functions, one more case must be considered, namely, integration between two discontinuities.

Consider the discontinuous function shown in Figure 5.6. We wish to integrate between a and b. We define the definite integral as in Eq. (5.35) with

$$\rightarrow\rightarrow\rightarrow \left\{ \begin{aligned} F(x)\Big|_a^b &= F(b-0) - F(a+0), \qquad a < b \\ F(x)\Big|_a^b &= F(b+0) - F(a-0), \qquad a > b \end{aligned} \right\} \tag{5.41}$$

A little reflection shows that Eqs. (5.41) hold whether or not f(x) is discontinuous at a and b. Also, the discontinuities can be infinite. This definition of definite integration is consistent with the rules of definite integration presented in standard analysis.

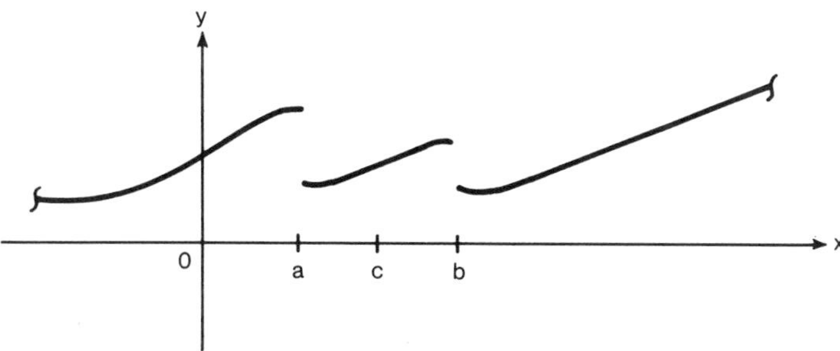

Figure 5.6 A generalized discontinuous function with two discontinuities.

An example of integration between discontinuities follows.

Example 5.10. Integration of $\text{saw}_\pi x$ between discontinuities and across two discontinuities:

$$\int_{-\pi/2}^{5\pi/2} \text{saw}_\pi x \, dx = \tfrac{1}{2} \, \text{saw}_\pi^2 x \Big|_{-\pi/2}^{5\pi/2}$$

$$= \tfrac{1}{2}(\tfrac{1}{2}\pi)^2 - \tfrac{1}{2}(-\tfrac{1}{2}\pi)^2 = 0$$

The developments in this section provide the basis for finding the definite integrals of broken and finitely-discontinuous functions over any interval containing any finite number of finite discontinuities and breaks in any combination, whether or not one or both limits of integration lie at a discontinuity or break. All of Section 5.4. can be summarized by noting that Eqs. (5.35) and (5.41) hold in general, with $f(x)$ being the complete derivative of $F(x)$.

5.5 The Definite Integral of a Staircase Function Is the Sum of a Series

A theorem concerning sums of series in terms of the definite integral of a staircase function is now presented.

Theorem 5.1

Given: $\sum_{n=a}^{b} f(n) = f(a) + f(a+1) + \cdots + f(b)$; a, b are integers, $a < b$.

$$\rightarrow\rightarrow\rightarrow \quad \int_{a-1/2}^{b+1/2} f[\sigma_1(x)]\,dx = \sum_{n=a}^{b} f(n)$$

where $\sigma_1(x) = x - \text{saw}_1 x$.

Proof of the theorem is given in Appendix A. Examples of the use of the theorem follow.

A trivial example is presented first, to get the reader into the spirit of the theorem. Let $f(n) = n$.

$$\int f[\sigma_1(x)]dx = \int \sigma_1(x)\,dx = \int x dx - \int \text{saw}_1 x dx. \tag{5.42}$$

The definite integral from $\frac{1}{2}$ to $b + \frac{1}{2}$ of this is

$$\sum_{n=1}^{b} n = \tfrac{1}{2}x^2 \Big|_{1/2}^{b+1/2} = \tfrac{1}{2}b^2 + \tfrac{1}{2}b. \tag{5.43}$$

This example is trivial because the term containing $\int \text{saw}_1 x dx$ drops out. The following two examples are nontrivial. This and the following examples can be extended to include the case $\sum_{n=a}^{b}$ by noting that

$$\sum_{n=a}^{b} f(n) = \sum_{n=1}^{b} f(n) - \sum_{n=1}^{a} f(n) + f(a).$$

We now examine the case $f(n) = n^2$. To solve this case we need the integrals of $\text{saw}_a^2 x$ and $x\ \text{saw}_a x$. These are found as follows.

We can obtain the integral of $\text{saw}_a^2 x$ by use of the Third Fundamental Theorem to obtain

$$\int \text{saw}_a^2 x\,dx = (1/3)\,\text{saw}_a^3 x + (a^2/12)\sigma_a(x) + C \tag{5.44}$$

We obtain the integral of $x\ \mathrm{saw}_a x$ by integrating by parts. It is

$$\int x\ \mathrm{saw}_a x\ dx = \tfrac{1}{2}x\ \mathrm{saw}_a^2 x - (1/6)\ \mathrm{saw}_a^3 x - (a^2/24)\sigma_a(x) + C \qquad (5.45)$$

We wish to find the integral of $\sigma_1^2(x)$. It is

$$\begin{aligned}\int \sigma_1^2(x)\ dx &= \int (x - \mathrm{saw}_1 x)^2 dx \\ &= (1/3)x^3 - x\ \mathrm{saw}_1^2 x + (2/3)\mathrm{saw}_1^3 x + (1/6)\sigma_1(x) + C\end{aligned} \qquad (5.46)$$

Finally we obtain

$$\sum_{n=1}^{b} n^2 = \int_{1/2}^{b+1/2} \sigma_1^2(x)\ dx = b^3/3 + b^2/2 + b/6 \qquad (5.47)$$

As a final example, we examine $f(n) = n^3$. For this case we need still more new integrals, which we now obtain.

By the Third Fundamental Theorem,

$$\int \mathrm{saw}_a^3 dx = \tfrac{1}{4}\ \mathrm{saw}_a^4 x + C \qquad (5.48)$$

Integrating by parts,

$$\begin{aligned}\int x\ \mathrm{saw}_a^2 x\ dx &= (a^2/24)x^2 - (a^2/12)x\ \mathrm{saw}_a x + (a^2/24)\mathrm{saw}_a^2 x \\ &= (1/3)x\ \mathrm{saw}_a^3 x - (1/12)\mathrm{saw}_a^4 x + C\end{aligned} \qquad (5.49)$$

$$\begin{aligned}\int x^2 \mathrm{saw}_a x\ dx = &- (a^2/24)x^2 + (a^2/12)x\ \mathrm{saw}_a x - (1/3)x\ \mathrm{saw}_a^3 x \\ &+ \tfrac{1}{2}x^2\ \mathrm{saw}_a^2 x - (a^2/24)\mathrm{saw}_a^2 x + (1/12)\mathrm{saw}_a^4 x + C\end{aligned} \qquad (5.50)$$

We wish to find the integral of $\sigma_1^3(x)$. It is

$$\begin{aligned}\int \sigma_1^3(x)dx &= \int (x - \mathrm{saw}_1 x)^3 dx \\ &= \tfrac{1}{4}x^4 + \tfrac{1}{4}x^2 - \tfrac{1}{2}x\ \mathrm{saw}_1 x - (3/2)x^2\mathrm{saw}_1^2 x + 2x\mathrm{saw}_1^3 x + \tfrac{1}{4}\mathrm{saw}_1^2 x \\ &\quad - (3/4)\mathrm{saw}_1^4 x + C\end{aligned} \qquad (5.51)$$

Finally we obtain

$$\sum_{n=1}^{b} n^3 = \int_{1/2}^{b+1/2} \sigma_1^3(x)\ dx = \tfrac{1}{4}b^4 + \tfrac{1}{2}b^3 + \tfrac{1}{4}b^2 \qquad (5.52)$$

5.6 Effective Value; rms

5.6.1 General

The root-mean-square or *rms* value of a periodic function of period a is, with b being an arbitrary constant,

$$y_{\text{rms}} = \left[\frac{1}{a} \int_{b-a/2}^{b+a/2} y^2(x)\,dx \right]^{1/2} \tag{5.53}$$

where $y^2(x) = [y(x)]^2$. The rms value is the *effective value* of an electrical current because the heating effect in a resistor is proportional to the square of the current. It is also the effective value of voltage because, again, the heating effect in a resistor is proportional to its square.

In the case of electrical power, P (the product of instantaneous voltage and current), the "effective" value of power[2] is just $\overline{P}$, the average value of the power function. Unfortunately, the term "effective value" has become synonymous with "rms value." We therefore find it necessary to introduce a new term: *power-equivalent value*. This is defined as the value of a constant quantity that signals the same power as the varying quantity. For current and voltage it is the rms value, but for power it is the average value.

In a manner similar to Eq. (5.53) we can define the root-mean-*n*th-power or rmn value as

$$y_{\text{rmn}} = \left[\frac{1}{\text{a}} \int_{b-a/2}^{b+a/2} y^n(x)\,dx \right]^{1/n} \tag{5.54}$$

We note that if n = 1, then $y_{\text{rmn}} = y_{\text{rm1}} = \overline{y}$.

If $y = y(x)$ is nonperiodic, as is the case with wind speed, then the rmn value is defined as[3]

$$y_{\text{rmn}} = \lim_{a\to\infty} \left[\frac{1}{a} \int_{b-a/2}^{b+a/2} y^n(x)\,dx \right]^{1/n} \tag{5.55}$$

The energy in wind is proportional to its speed squared and power is the time rate of flow of energy; thus wind power is proportional to wind speed cubed. Therefore, the power-equivalent value of wind speed is its rm3 value: its "root-mean-cube." This expresses its potential capability to drive a wind-powered machine.[4]

[2] Again, in terms of its heating effect in a resistor.

[3] The constant b is usually taken as zero in this case.

[4] See also H. C. Kelly, "Renewable Energy," *Research and Development*, Vol. 26, No. 6, June 1984, pp. 213–220.

The term "true rms" is also encountered. This term is necessitated by the fact that many so-called rms voltmeters read rms only for sinusoidal waves. The term *true rms* is then used to refer to voltmeters that read rms for an arbitrary waveform.

Example. The rms value of $y = A$ tri x is

$$[A \text{ tri } x]_{\text{rms}} = A\pi 3^{-1/2}$$

5.6.2 Electronic implementation of the rms operation

The rms value of a periodic function p of period a can be found by use of the averaging operation $\overline{(\ )}$ in the expression

$$p_{\text{rms}} = [\overline{(p^2)}]^{1/2} \tag{5.56}$$

It is obvious that p_{rms} can never be negative. Equation (5.56) is diagrammed in Fig. 5.7. This method can be used for a range of rms values that extends all the way down to zero. For rms values that never get closer to zero than a particular value e, other methods are possible, as is shown next.

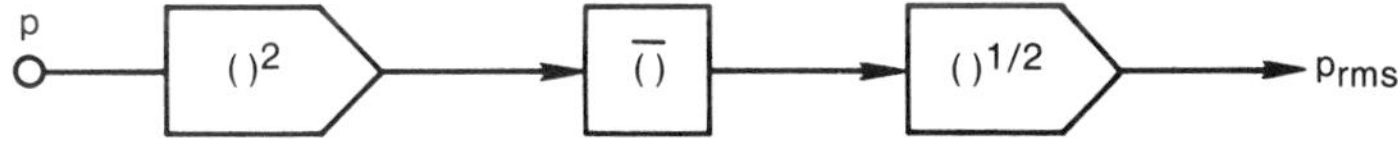

Figure 5.7 Block diagram of the rms operation of Eq. (5.56).

From Eq. (5.56) we can write

$$p_{\text{rms}} = \overline{(p^3)}/p_{\text{rms}} \tag{5.57}$$

Equation (5.57) can be solved implicitly as shown in Fig. 5.8. This allows the square-root operation to be replaced by division, seemingly not much of an advantage. But if p is truly periodic or stationary (of constant amplitude, frequency, and waveshape), then p_{rms} will be constant; and Eq. (5.57) can be written

$$p_{\text{rms}} = \overline{u} \tag{5.58}$$

where

$$u = p^2/p_{\text{rms}} \tag{5.59}$$

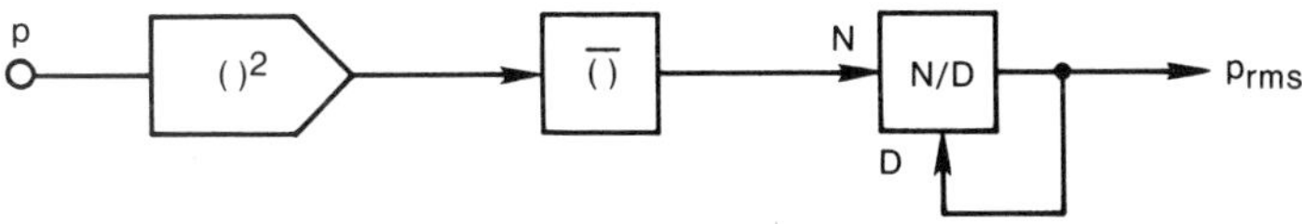

Figure 5.8 Block diagram of the rms operation of Eq. (5.57).

This has the effect of allowing the last two operations of Fig. 5.8 to be interchanged; the advantage of this is that squaring and division operations can be performed by a single multiplier/divider—a readily available electronic module or integrated circuit. This method is diagrammed in Fig. 5.9.

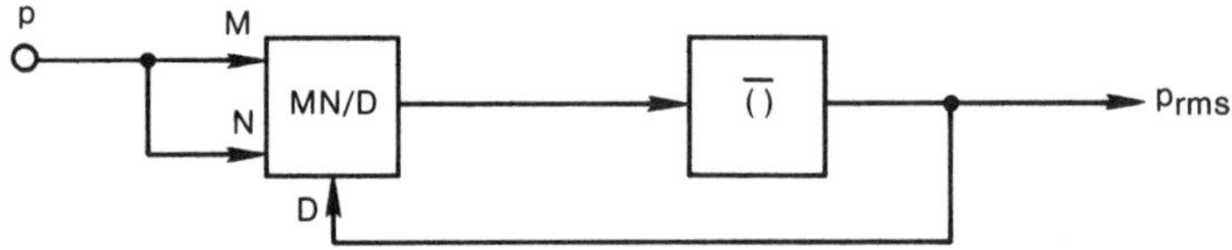

Figure 5.9 Block diagram of the rms operation of Eq. (5.58).

For functions of constant and known waveshape, further simplifications can be had. If the d-c component is zero, one needs only to detect the peak value (amplitude) or the average of $|p|$ and scale that quantity to obtain the rms value. This is the classical method used in so-called rms voltmeters (as opposed to true rms voltmeters) that assume a sinewave input.

This latter method has an advantage over the methods of Fig. 5.6 and 5.7 in that it is usable for rms values all the way to zero.

A practical device based on this method that is usable for different waveforms is diagrammed in Fig. 5.10. The first block *A* detects some convenient property of *p* such as amplitude. This is followed by a manual selector that is set for the particular waveform being measured.

If amplitude is detected, this device has a further advantage: integration (averaging) with its attendant time-constant considerations is avoided.

The preceding paragraphs address only periodic functions. Therefore noise signals—since they are not periodic—are not included.

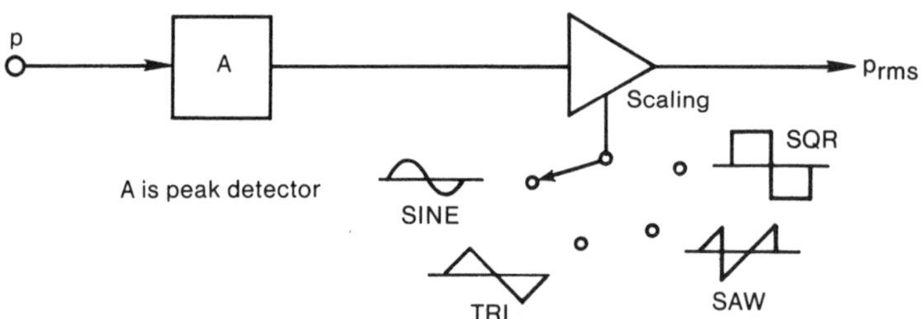

Figure 5.10 Block diagram of the rms operation when waveform is known.

For nonperiodic functions, the methods of Fig. 5.7 and Fig. 5.8 can be used. The difficulty in this case is the choice of time constant for the averaging operation. In the implementations, some compromise in this regard must usually be made because use of the ideal time constant (infinity) is out of the question when p is a function of time. One usually tries to select a value for the time constant that is large compared to the period of the lowest-frequency sinewave component of interest in the signal. How large depends on the accuracy desired.

5.7 Summary

Many of the integrals of this chapter are of the form $\int \text{stair}(x)f(x)dx$ where $\text{stair}(x)$ is a zero-slope function. Among these are $\int \text{sgn}x f(x)dx$, $\int \text{step}(x - a)f(x)dx$, $\int |\sin x|dx = \int \text{sqr}\, x \sin x\, dx$, $\int |\cos x|dx = \int \text{sgn}(\cos x) \cos x\, dx$, and $\int f[\sigma_1(x)]dx$. A theorem concerning this general form is presented as Theorem 5.2.

Theorem 5.2

Given: The function $g(x) = \text{stair}(x)f(x)$ is no more than finitely discontinuous.

$$\rightarrow\rightarrow\rightarrow \quad \int g(x)dx = \text{stair}(x)\int f(x)dx + \text{stair}'x.$$

Where: **1.** $\text{stair}(x)$ and $\text{stair}'(x)$ are zero-slope functions of x,
2. $\text{stair}'(x)$ is such that $\int g(x)dx$ is continuous.

This theorem is an integral form of Theorem 4.1. By "no more than finitely discontinuous" is meant it is continuous throughout except for possibly one or more isolated finite discontinuities.

It is noted that a zero-slope function of a function of x is itself a zero-slope function of x, so that $\text{stair}[u(x)] = \text{stair}'(x)$; and a function of a zero-slope function is itself a zero-slope function, so that $f(\text{stair}\ x) = \text{stair}'(x)$.

In the following chapter we take a close look at wide pulses and their applications. These pulses were introduced in Chapter 2.

6

WIDE PULSES AND THEIR APPLICATIONS

6.1 How to Synthesize Rectangular Pulse Functions and Pulse Trains

Pulses that have a finite width are said to be *wide*. This is to distinguish them from delta pulses. Wide pulse functions were introduced in Chapter 2 as combinations of step functions. Cf. Eq. (2.25), repeated here:

$$\text{puls}_a^b x \equiv S(x - a) - S(x - b)$$

Another way to synthesize pulse functions is by making use of the fact that $S(u)$ is a function of unit rectangular pulses in the variable x when $u = u(x)$. For example, $S(1 - |x|)$ is the pulse function graphed in Figure 6.1. This is the function $\text{puls}_{-1}^1 x$. Another example of this kind of *pulse-forming operation* was given in Chapter 2 in Eq. (2.44) as the square pulse train $S(\sin x)$. In general, $S(u) = 1$ when $u > 0$ and $S(u) = 0$ when $u < 0$. When $u = 0$, $S(u)$ is 0 or 1 depending on its immediately preceding value for $u \neq 0$. This action is due to the operation of infinitesimal hysteresis.

The functions $S(\sin x)$ and $S(\cos x)$ belong to the general class of *rectangular pulse trains*. Specifically, they are *square pulse trains* because the pulse width equals half the fundamental period. In general, the ratio of pulse width to pulse period is the *duty ratio,* usually expressed as a percent. Duty ratio is also called "duty factor" or "duty cycle." The ratio of pulse width to period minus pulse width is called *mark/space ratio*—a term from telegraphy.

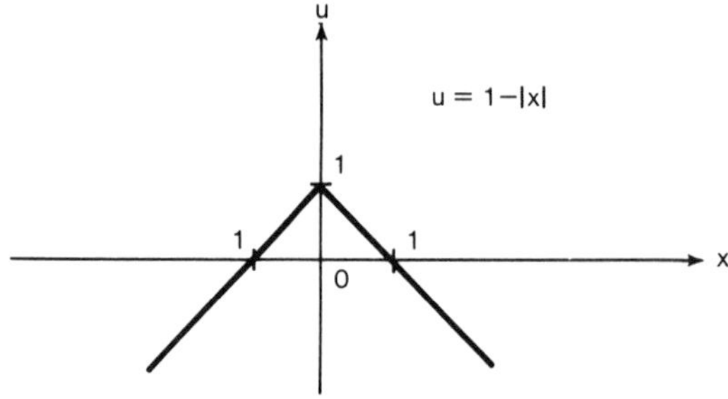

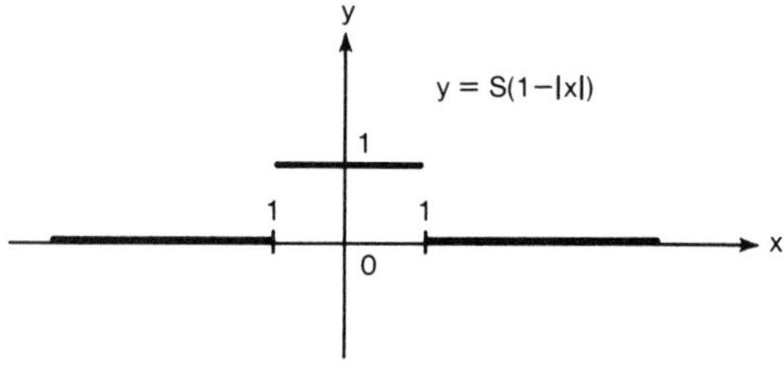

Figure 6.1 A pulse function formed by use of the Heaviside step operation.

Consider the function

$$y = S(b\pi - \text{tri}\, x) \tag{6.1}$$

This is a train of rectangular pulses with duty ratio b $(0 < b < 1)$ and period 2π. Finally, the function

$$y = \text{sgn}(\cos x)S(b\pi - \tfrac{1}{2}\text{tri}\, 2x) \tag{6.2}$$

is a train of *alternating* rectangular pulses with duty ratio b $(0 < b < \frac{1}{2})$ and period 2π.

The pulse function of Eq. (2.25) can also be written

$$\text{puls}_a^b x = \tfrac{1}{2}\text{sgn}(x - a) - \tfrac{1}{2}\text{sgn}(x - b) \tag{6.3}$$

The integral of this can readily be obtained as

$$\int \text{puls}_a^b x\, dx = \tfrac{1}{2}|x - a| - \tfrac{1}{2}|x - b| + C \tag{6.4}$$

This is the *linear ramp function* graphed in Fig. 6.2. The height of the ramp is $b - a$. Its slope is equal to the amplitude of the pulse that is its derivative, in this case $+1$ for $a < b$ and -1 for $a > b$.

As with delta pulses, we can speak of the *content* of wide pulses. The content of a wide pulse is the height of the ramp that is formed upon its integration. This is also equal to the area under the pulse. This definition is consistent with the definition of the content of a delta pulse because a step function can be considered to be a limiting case of a ramp.

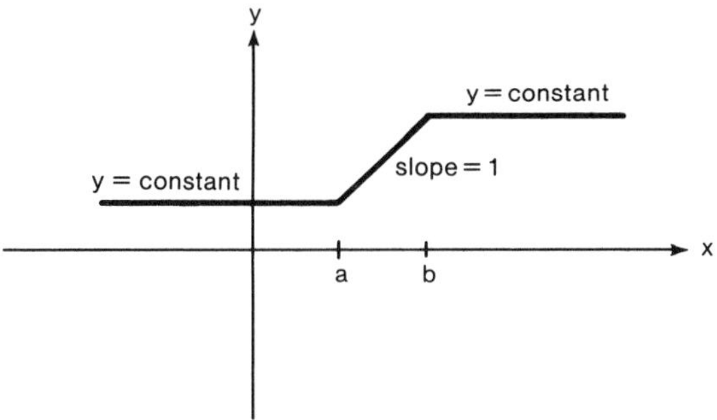

Figure 6.2 The linear ramp function $y = \frac{1}{2}|x - a| - \frac{1}{2}|x - b|$.

The derivative of *puls* is

$$\frac{d}{dx}\text{puls}_a^b x = \delta(x - a) - \delta(x - b) \tag{6.5}$$

Other relationships are

$$\text{puls}_b^a x = -\text{puls}_a^b x \tag{6.6}$$

$$\text{puls}_a^b(-x) = -\text{puls}_{-a}^{-b} x \tag{6.7}$$

The *pulse-width modulation* operation can be expressed as $S(u - v)$ where u is a triangular wave of high frequency and v is the modulation. We can write

$$u = \text{tri}(\omega_1 t) \tag{6.8a}$$

$$v = \sin(\omega_2 t) + \tfrac{1}{2}\pi, \qquad \omega_1 >> \omega_2 \tag{6.8b}$$

Thus $S(u - v)$ is a train of pulses of angular frequency ω_1 whose duty ratio depends on the modulating signal. Oxner[1] discusses this operation.

Pulse-width modulation finds application in communications, switching power supplies, motor control, and switch-mode audio amplifiers.

6.2 Using Rectangular Pulse Functions as Gates

A rectangular pulse can be used as a *gate* to *enable* an arbitrary function within the interval of the pulse and to *disable* it elsewhere. This can be written

$$y = f(x)\,\text{puls}_a^b x \tag{6.9}$$

It is clear that Eq. (6.9) means, if $a < b$,

$$\begin{aligned} y &= f(x) \qquad \text{when } a \underset{\leftarrow}{\leq} x \underset{\rightarrow}{\leq} b \\ &= 0 \text{ elsewhere} \end{aligned} \tag{6.10}$$

[1] See pages 205 and 139 of E. S. Oxner, *Power FETs and Their Applications* (Englewood Cliffs, N.J.: Prentice-Hall, Inc., 1982).

Figure 6.3 shows gated sinewaves—"wavetrains"—of one and two cycles, where *gating-on* and *gating-off* occur at points where $\sin x = 0$. An equation for this kind of gating over N cycles, starting at $x = 0$, is

$$y = \sin x \operatorname{puls}_0^{2N\pi}(x) \tag{6.11}$$

The integral and derivative of Eq. (6.11) are

$$\int y\,dx = -\cos x \operatorname{puls}_0^{2N\pi}(x) + C \tag{6.12}$$

$$\frac{dy}{dx} = \cos x \operatorname{puls}_0^{2N\pi}(x) \tag{6.13}$$

When gating-on and gating-off occur at points where $\sin x \neq 0$, a delta function is present in the derivative. The integral and derivative in the general case are

$$\int f(x)\operatorname{puls}_a^b x\,dx = \operatorname{puls}_a^b x \int f(x)dx + \left[S(x-u)\int f(u)du\right]_{u=a}^{u=b} + C \tag{6.14}$$

$$\frac{d}{dx}[f(x)\operatorname{puls}_a^b x] = \operatorname{puls}_a^b x \frac{d}{dx}f(x) + f(x)[\delta(x-a) - \delta(x-b)] \tag{6.15}$$

It is not difficult to show that the definite integral over all x-values of a general gated function is

$$\int_{-\infty}^{\infty} f(x)\operatorname{puls}_a^b x\,dx = \int_a^b f(x)dx \tag{6.16}$$

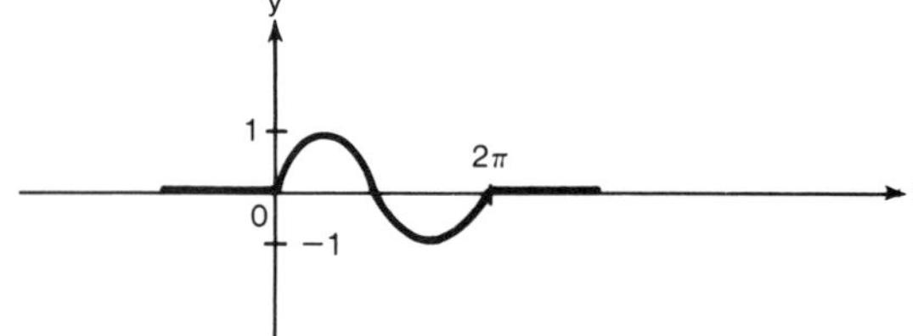

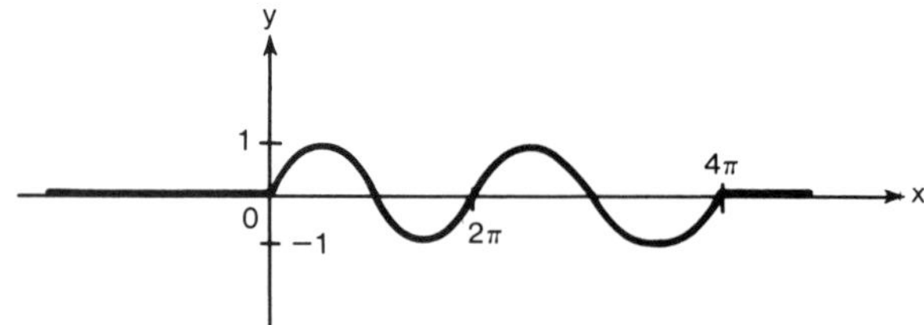

Figure 6.3 Gated sinewaves of one and two cycles.

6.3 How to Splice Functions to Produce New Functions

The First Fundamental Theorem facilitates the synthesis of virtually any desired periodic function. *Function splicing* is a procedure that facilitates the synthesis of virtually any desired nonperiodic function.

Function splicing is the joining of two or more single-valued functions to form a new single-valued function. We desire, with $a < b < c$,

$$\left.\begin{array}{ll} y = f_1(x) & \text{when } a \underset{\leftarrow}{\le} x \underset{\rightarrow}{\le} b \\ y = f_2(x) & \text{when } b \underset{\leftarrow}{\le} x \underset{\rightarrow}{\le} c \\ y = 0 & \text{elsewhere} \end{array}\right\} \tag{6.17}$$

We can write the function of Eqs. (6.17) as a sum of gated functions as

$$y = f_1(x)\text{puls}_a^b x + f_2(x)\text{puls}_b^c x \tag{6.18}$$

As an example of the use of Eq. (6.18), suppose we want to synthesize the nonperiodic *triangular pulse function* shown in Fig. 6.4. The desired expression is readily seen to be

$$y = (x - 1)\text{puls}_1^2 x + (-x + 3)\text{puls}_2^3 x \tag{6.19}$$

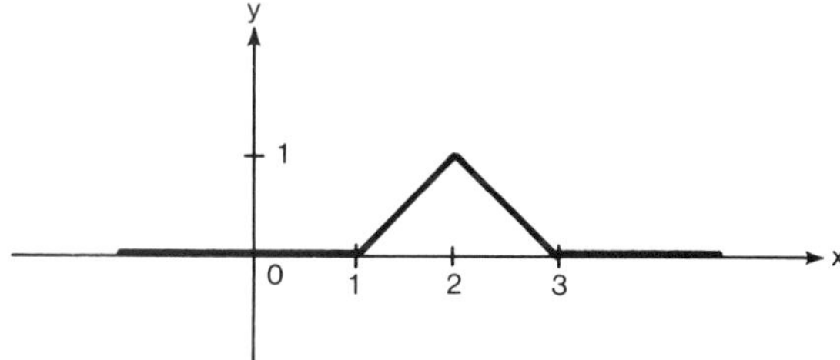

Figure 6.4 A triangular pulse function.

In Eq. (6.18), the point at the splice is at $y = f_1(b)$ for positive progression and at $y = f_2(b)$ for negative progression. It follows that the point at the splice in Eq. (6.19) is at $y = 1$ for both directions of progression.

The processes of gating and function splicing can be instrumented by the use of *analog switches,* discussed later. Other applications of function splicing are considered next.

6.4 Sample-and-Hold and Related Operations

The *sample-and-hold* or *track-and-hold* operation converts a function $f(x)$ into another function $g(x)$ such that

$$g(x) = f(x), \qquad x < a \tag{6.20a}$$

$$g(x) = f(a), \qquad x \geqslant a \tag{6.20b}$$

This operation can be performed by using *semi-infinite pulse functions* (step functions) as gates in a function-splicing operation.

Equations (6.20a) and (6.20b) can be combined to give

$$g(x) = f(x)S(-x + a) + f(a)S(x - a) \tag{6.21}$$

Operations which, at first glance, appear to be related to the sample-and-hold operation are *peak detection* and *minimum detection*. Indeed, these operations can be performed by multiple function splicing; however, they are more readily performed in another manner.

The operations of *peak detection, maximum detection,* and *minimum detection* will be referred to collectively as the *monotonic operation*. The output of the monotonic operation is a function that is either nondecreasing or nonincreasing.

For example, if the input function $u = u(x)$ is a function such as the one shown in Fig. 6.5a, then the output function $y = f(x)$ is either the function shown in Fig. 6.5b or the one in Fig. 6.5c. This operation can be described by the differential equation

$$dy/dx = S(du/dx)du/dx \tag{6.22}$$

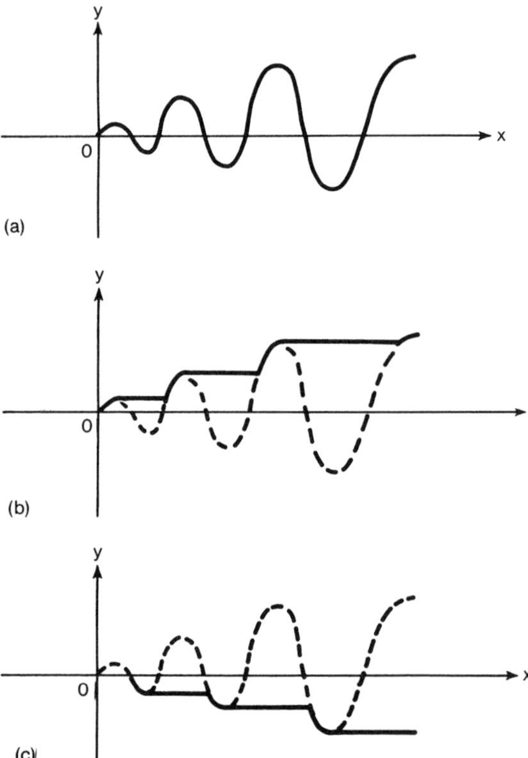

Figure 6.5 An illustration of the monotonic operation: (a) an "arbitrary" input function; (b) the output function for the non-decreasing monotonic operation; (c) the output function for the non-increasing monotonic operation.

for the *nondecreasing* monotonic operation or by

$$dy/dx = S(-du/dx)du/dx \tag{6.23}$$

for the *nonincreasing* monotonic operation. Thus in Eq. (6.22), when du/dx is positive, dy/dx tracks it; when du/dx is negative, $dy/dx = 0$. In addition, the condition $y = u$ when $S(\ ) = 1$ also holds. If $u = u(x)$ is continuous, then $y = y(x)$ is also continuous.

Equations (6.22) and (6.23) can be integrated by use of Theorem 5.2. Both integrals are of the form

$$y = \int \text{stair}\ x\ du \tag{6.24}$$

because $S(du/dx)$ and $S(-du/dx)$ are both zero-slope functions of x. Therefore, we can integrate Eq. (6.22) to obtain

$$y = \int S(du/dx)du = uS(du/dx) + \text{stair}\ x \tag{6.25}$$

for the nondecreasing monotonic operation. (The constant of integration is contained in stair x.) Similarly,

$$y = uS(-du/dx) + \text{stair}\ x \tag{6.26}$$

for the nonincreasing monotonic operation. We want $y = u$ when $S(\ ) = 1$, so the zero slope functions must be zero in those intervals.

6.5 Analog Switches and Other Quasi-Multipliers

Multiplication is a basic mathematical operation that can be performed by an electronic analog multiplier. An operation that we shall call *quasi-multiplication,* while not differing from ordinary multiplication in the mathematical sense, is special in that its electronic implementation can be performed by devices that are substantially simpler and less expensive than multipliers. We devote this section to a discussion of quasi-multiplication and devices that perform it.

If one of the functions involved in a multiplication is a zero-slope function, then the operation is termed *quasi-multiplication* and it can be performed by a *quasi-multiplier.* Four kinds of quasi-multipliers will be discussed: *analog switches,* (*analog*) *multiplexers, programmable-gain devices,* and *AND-gates.*

Analog switches perform the multiplication $f(x)S(u)$. They have a *signal* (or *analog*) *input* that accepts an arbitrary function $f(x)$, a *control* (or *digital*) *input* that accepts a step function $S(u)$, and an output that is the product of the two inputs.

Analog switches are very fast solid-state relays whose "contacts" are either closed or open depending on whether the control input is ZERO or ONE. If the switch is ON ("contacts" closed) when $S = 0$ it is said to be *normally closed;* if it is OFF ("contacts" open) when $S = 0$ it is said to be *normally open.*

If $S(u) = \text{puls}_a^b x$ then the operation $f(x)S(u)$ that is performed by a normally open analog switch is simply the gating operation. In addition to gating, analog switches can perform *polarity selection.* This can be expressed as

$$y = f(x)S(u) - f(x)S(-u) \tag{6.27}$$

where $S(u)$ represents a normally open device and $S(-u) = 1 - S(u)$ represents a normally closed device. A circuit that performs this operation is shown in Fig. 6.6. The output is either $f(x)$ or $-f(x)$ depending on the control input.

Analog switches with proper inputs can also be used to perform full-wave rectification, half-wave rectification, and pseudorectification (to produce pseudorectified sine and cosine waves) by use of gating and polarity selection operations. Block diagrams illustrating the use of analog switches for performing full- and half-wave rectification are developed in Chapter 7. Analog switches are also called *analog gates.*

Multiplexers are essentially multiple analog switches incorporating binary digital logic stages to reduce the number of control inputs required. The inputs to these logic stages are called *addresses.* In this way if eight

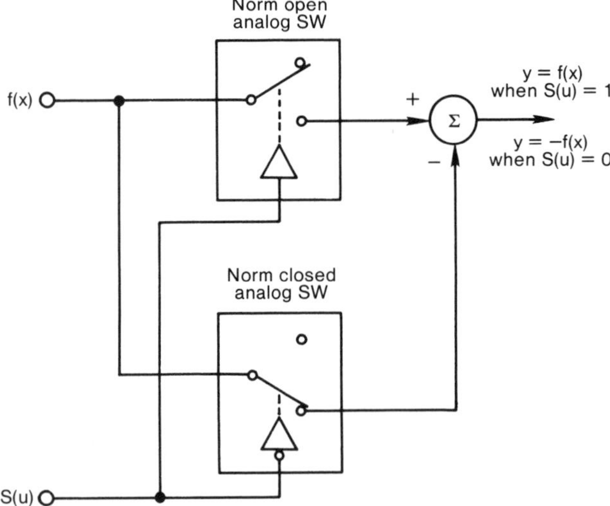

Figure 6.6 Block diagram of the polarity selection operation.

switches are to be controlled, for example, only three address inputs are required. The switches themselves are normally open, and only one switch can be closed at a time.

Programmable-gain devices perform the multiplication $f(x)$ stair(u) where stair(u) is virtually any desired staircase function. They can take the form of an operational amplifier with analog switches or multiplexers controlling the value of the feedback resistor.

AND-gates perform the multiplication $S(u_1)S(u_2)$. Since the two inputs are symmetrical (they can both be thought of as control inputs), it is easy to extend this concept to more than two inputs. Such a device performs the multiplication $S(u_1)S(u_2) \ldots S(u_n)$ where n is the number of inputs. It is easy to see that regardless of the state of the inputs (whether ZERO or ONE), the output can take on only the values 0 and 1. A three-input AND-gate is represented by the block diagram in Fig. 6.7. The AND-gate is a true *binary digital logic element* or device. We find use for it in an analog application in Chapter 10.

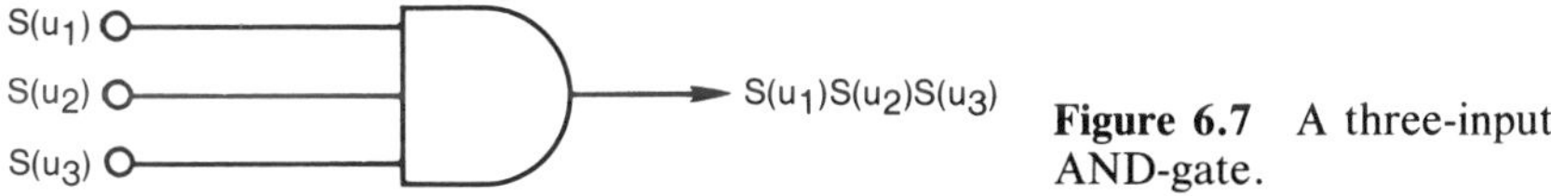

Figure 6.7 A three-input AND-gate.

Besides being simpler and therefore less expensive than multipliers, analog switches and other quasi-multipliers do not require an exact control input to do accurate quasi-multiplication. For example, typical specifications for an analog switch (using bipolar technology) permit the control input for ZERO to be anywhere between -5.0 and $+0.8$ volts, and that for ONE to be anywhere between 2.4 and 5.0 volts. This is also the situation for all inputs of an AND-gate. By the same token, AND-gates are designed to drive other digital devices and so one cannot depend on their outputs to be precise in an analog application.

6.6 Zero-Crossing Detection

A property of the signum operation is that, for a function $u = u(x)$, sgn u is a function of *bipolar rectangular pulses* in the variable x. When $u > 0$, sgn $u = 1$, and when $u < 0$, sgn $u = -1$. Because of infinitesimal hysteresis, sgn $0 = \pm 1$, depending on its immediately-preceding value for $u \neq 0$.

Since sgn u is constant in magnitude but changes sign when u crosses zero, application of the signum operation constitutes *zero-crossing detection* for continuous u-functions.[2] As an example, sqr x is the *zero-crossing function* of sin x.

Consequently, zero-crossing detection can be performed by a comparator. It can also be approximated by a process known as "infinite clipping," discussed in Chapter 7.

Zero-crossing detection can be used to produce bipolar rectangular pulse functions and pulse trains just as the pulse forming operation, $S(u)$, is used to produce positive ones.

A second kind of zero-crossing detection is discussed by Meiksin and Thackray.[3] In this kind, a positive pulse is produced for each zero crossing. A similar kind of pulse-forming operation is discussed herein in Section 7.6.3.

Giles[4] gives a practical circuit for the first kind of zero-crossing detection.

6.7 Sinusoidal Pulses

In addition to the rectangular and triangular pulses of preceding sections, sinusoidal pulses such as the one shown in Fig. 6.8 are of interest in the applications. The equation of the pulse function of Fig. 6.8 is

$$y = \cos x \, \text{puls}_{-\pi/2}^{\pi/2}(x) \tag{6.28}$$

The width of this pulse at its base is π. In general we can write for a sinusoidal pulse function with pulse of base width $2a$ $(0 < a < \frac{1}{2}\pi)$

$$y = A \cos \frac{\pi x}{2a} \text{puls}_{-a}^{a} x \tag{6.29}$$

The amplitude of this pulse is A.

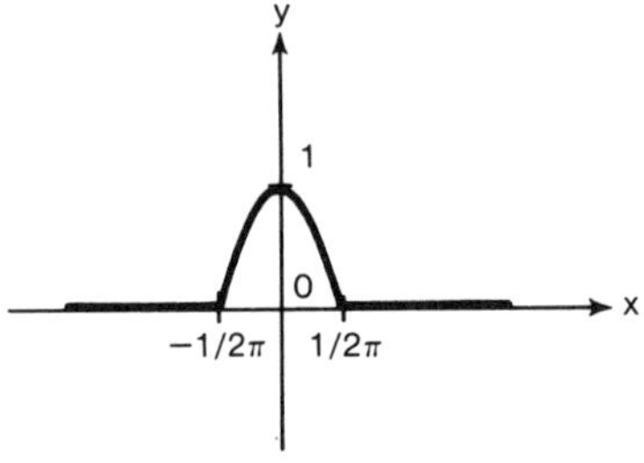

Figure 6.8 A sinusoidal pulse function.

[2]A discontinuous function can change sign without literally crossing zero, as $\text{saw}_a x$ does at $x = \frac{1}{2}a$. We restrict use of the term "zero-crossing detection" to continuous u-functions.

[3]Z. H. Meiksin and P. C. Thackray, *Electronic Design with Off-the-Shelf Integrated Circuits*, 2nd ed. (Englewood Cliffs, N.J.: Prentice-Hall, Inc., 1984).

[4]J. N. Giles, *Fairchild Semiconductor Linear Integrated Circuits Applications Handbook*, Fairchild Semiconductor, 313 Fairchild Drive, Mountain View, Cal., 1967.

If we replace x with $\text{saw}_\pi x$ in Eq. (6.29), then the equation describes the sinusoidal pulse train graphed in Fig. 6.9. Similarly,

$$y = A\cos\frac{\pi u}{2a}\text{puls}^{a}_{-a}u\ \text{sgn}(\cos x) \tag{6.30}$$

with $u = \text{saw}_\pi x$, describes the alternating sinusoidal pulse train graphed in Fig. 6.10. This pulse train is characteristic of the tank current in a ferroresonant power supply.[5]

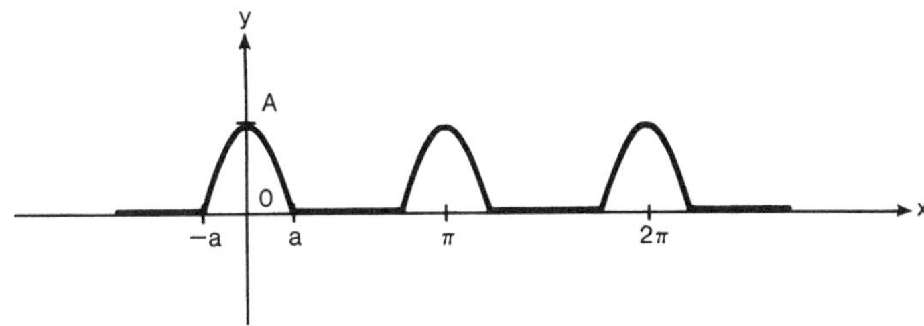

Figure 6.9 A sinusoidal pulse train.

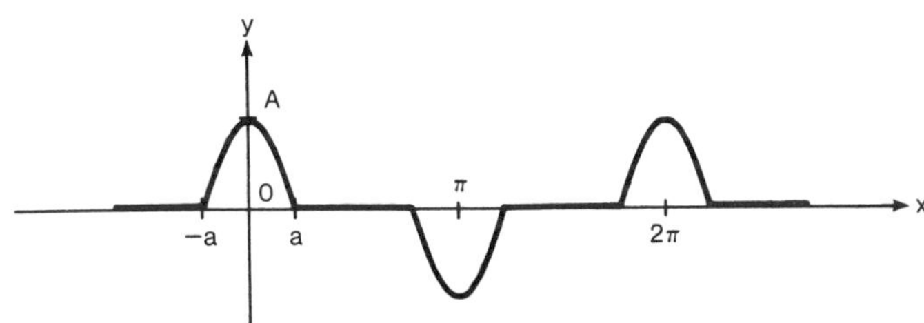

Figure 6.10 An alternating sinusoidal pulse train.

The integral of the sinusoidal pulse function of Eq. (6.29) is the *sinusoidal ramp function*

$$\int y\,dx = A(2a/\pi)\left[\sin\frac{\pi x}{2a}\text{puls}^{a}_{-a}x + S(x+a) + S(x-a) - 1\right] + C \tag{6.31}$$

A graph of this function with $C = 0$ is shown in Fig. 6.11.

[5]See, for example, page 122 of *Kepco Power Supply Catalog & Handbook,* no. 146–1402, Kepco, Inc., 1981.

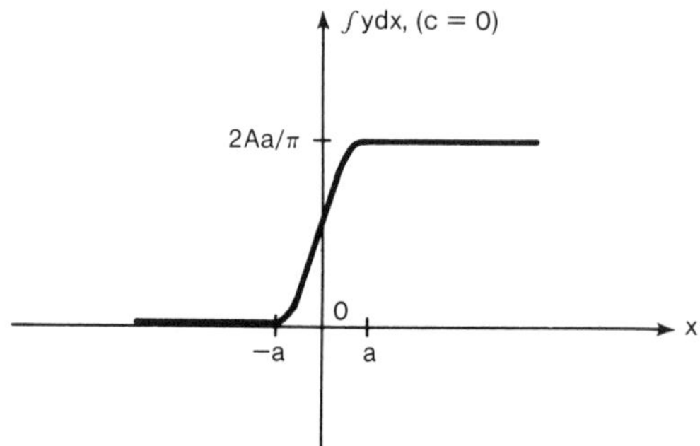

Figure 6.11 A sinusoidal ramp function.

The sinusoidal pulse functions of Eqs. (6.29) and (6.30) are continuous-broken functions. The function

$$y = (\cos x + 1)\text{puls}_{-\pi}^{\pi} x \tag{6.32}$$

is the continuous-smooth sinusoidal pulse function graphed in Fig. 6.12. Its derivative is

$$\frac{dy}{dx} = -\sin x \; \text{puls}_{-\pi}^{\pi} x \tag{6.33}$$

This is the continuous-broken *bipolar* sinusoidal pulse function graphed in Fig. 6.13.

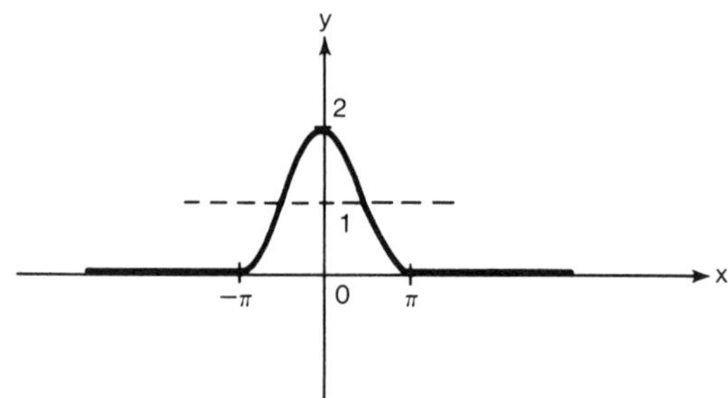

Figure 6.12 A continuous-smooth sinusoidal pulse function.

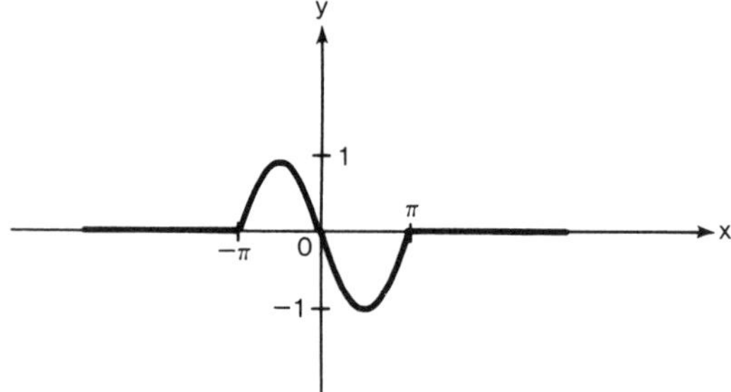

Figure 6.13 The derivative of the pulse function of Fig. 6.12.

6.8 Overview

The wide pulse functions, gates, ramp functions, and zero-crossing detection discussed in this chapter are closely related to clipping. For example, the clipping operation is the integral of the pulse-forming operation *puls*. This and other important relationships form the subject matter of the next chapter.

7

CLIPPING AND RELATED OPERATIONS

7.1 Half-Wave Rectification Is Clipping at a Lower Level of Zero

7.1.1 Half-wave rectification

If the function of unit pulses $S(u)$ is used to gate u itself, the result is the *half-wave rectification* of u, $\mathrm{rec}_{1/2}u$. We define

$$\rightarrow\rightarrow\rightarrow \quad \mathrm{rec}_{1/2}u \equiv uS(u) \tag{7.1}$$

Since $S(u) = \frac{1}{2} + \frac{1}{2}\mathrm{sgn}\, u$, we can also write

$$\mathrm{rec}_{1/2}u = \tfrac{1}{2}u + \tfrac{1}{2}|u| \tag{7.2}$$

The half-wave rectified sinewave can be written

$$\mathrm{rec}_{1/2}(\sin x) = \sin xS(\sin x) = \tfrac{1}{2}\sin x + \tfrac{1}{2}|\sin x| \tag{7.3}$$

A graph of $y = \mathrm{rec}_{1/2}u$ is shown in Fig. 7.1. A graph of $\mathrm{rec}_{1/2}(\sin x)$ appears in Fig. 1.3 of Chapter 1.

In a similar way, if the function of unit pulses $S(-u)$ is used to gate u, the *negative half-wave rectification* of u, $\mathrm{rec}_{-1/2}u$, results. A graph of $\mathrm{rec}_{-1/2}u$ appears in Fig. 7.2. This is a negative form of half-wave rectification. It eliminates the positive portions of u just as $\mathrm{rec}_{1/2}u$ eliminates the negative portions. We define

$$\mathrm{rec}_{-1/2}u \equiv uS(-u) \tag{7.4}$$

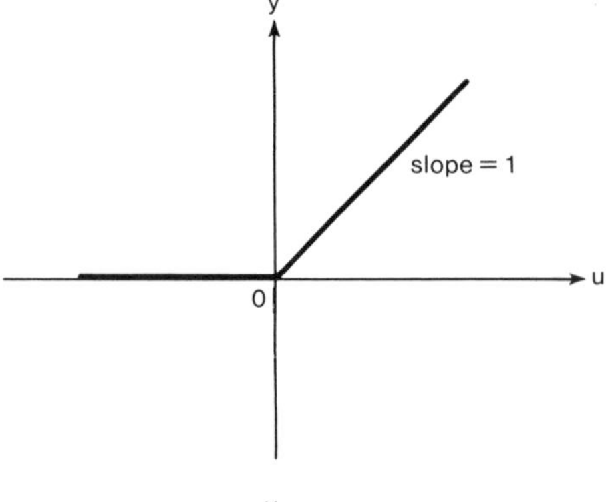

Figure 7.1 The half-wave rectification operation.

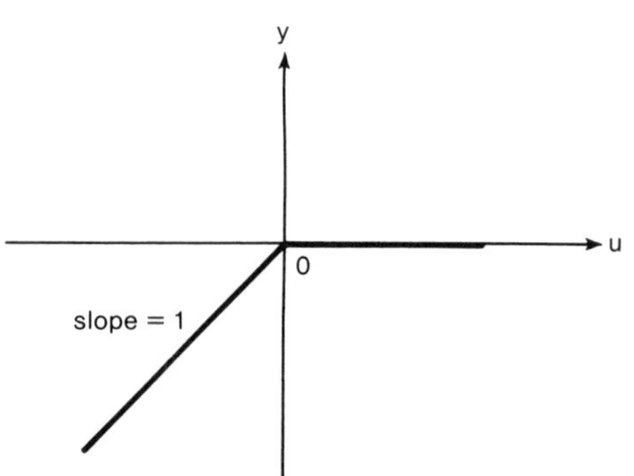

Figure 7.2 The negative half-wave rectification operation.

We can also write

$$\text{rec}_{-1/2}u = \tfrac{1}{2}u - \tfrac{1}{2}|u| \tag{7.5}$$

$$\text{rec}_{-1/2}u = u - \text{rec}_{1/2}u = -\text{rec}_{1/2}(-u) \tag{7.6}$$

$$\text{rec}_{1/2}u - \text{rec}_{-1/2}u = |u| \tag{7.7}$$

This last equation shows how full-wave rectification can be performed by combining the two forms of half-wave rectification, $\text{rec}_{1/2}$ and $\text{rec}_{-1/2}$. It can easily be seen that full-wave rectification can also be performed by the use of $\text{rec}_{1/2}$ alone or $\text{rec}_{-1/2}$ alone, because

$$|u| = 2\,\text{rec}_{1/2}u - u \tag{7.8}$$

$$|u| = -2\,\text{rec}_{-1/2}u + u \tag{7.9}$$

from Eqs. (7.2) and (7.5), respectively.

The derivatives and integrals of $\text{rec}_{1/2}$ and $\text{rec}_{-1/2}$ can be obtained by starting with Eqs. (7.1) or (7.2) and (7.4) or (7.5). Examples follow.

Example 7.1. The derivative of $\text{rec}_{1/2}\sin x$ is

$$\begin{aligned}\frac{d}{dx}\,\text{rec}_{1/2}\sin x &= \frac{d}{dx}\sin xS(\sin x)\\ &= \cos xS(\sin x)\end{aligned}$$

Example 7.2. The derivative of $\text{rec}_{-1/2}\sin x$ is

$$\frac{d}{dx}\,\text{rec}_{-1/2}\sin x = \frac{d}{dx}\sin xS(-\sin x)$$
$$= \cos xS(-\sin x)$$

Example 7.3. The integral of $\text{rec}_{1/2}\sin x$ is

$$\int \text{rec}_{1/2}\sin x\,dx = \int \tfrac{1}{2}\sin x\,dx + \int \tfrac{1}{2}|\sin x|\,dx$$
$$= -\tfrac{1}{2}\cos x + \tfrac{1}{2}\int |\sin x|\,dx, \text{ where } \int |\sin x|\,dx$$

is given by Eq. (5.32)

The term "half-wave rectification" is an electronic term. There is no equivalent mathematical term for this operation.

7.1.2 All-wave rectification

We have had occasion to introduce three kinds of rectification: full wave, half wave, and negative half wave. Each of these can be written in the form

$$\text{rec}\,u = u\,\text{step}\,u \tag{7.10}$$

where

$$\text{step}\,u = \text{sgn}\,u \qquad \text{when} \qquad \text{rec}\,u = |u| \tag{7.11a}$$
$$\text{step}\,u = S(u) \qquad \text{when} \qquad \text{rec}\,u = \text{rec}_{1/2}u \tag{7.11b}$$
$$\text{step}\,u = S(-u) \qquad \text{when} \qquad \text{rec}\,u = \text{rec}_{-1/2}u \tag{7.11c}$$

If we agree to write $\text{rec}_1 u$ for $|u|$ then we can define a generalized *all-wave rectification*, rec_w, which encompasses full-wave, half-wave, and negative half-wave rectification as special cases. We define

$$\rightarrow\rightarrow\rightarrow \quad \text{rec}_w u \equiv u\,\text{step}_w u \tag{7.12}$$

where step_w means st $\underline{1 - |w| - w}\overline{1 - |w| + w}$ (cf. Section 2.2), so that

$$\text{step}_w u = 1 - |w| + w\,\text{sgn}\,u \tag{7.13}$$

and

$$\rightarrow\rightarrow\rightarrow \quad \text{rec}_w u = u - u\,|w| + |u|w \tag{7.14}$$

It can be seen that

$$\text{rec}_w(-u) = -\text{rec}_{-w}u \tag{7.15}$$

Graphs of $\text{step}_w x$ for various values of w are shown in Fig. 7.3.

Note that w need not be restricted to the range -1 to $+1$, thereby

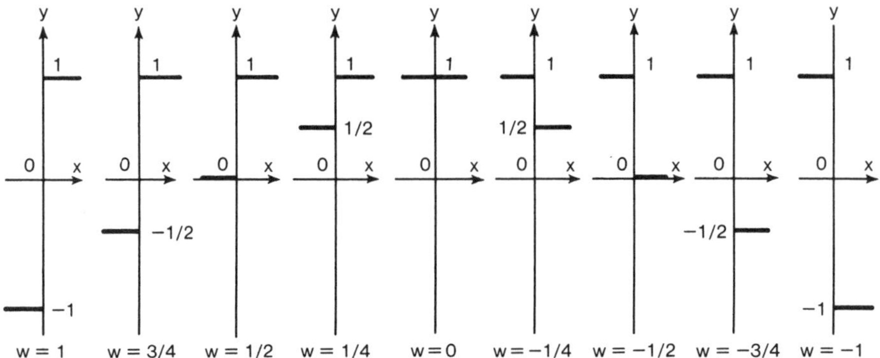

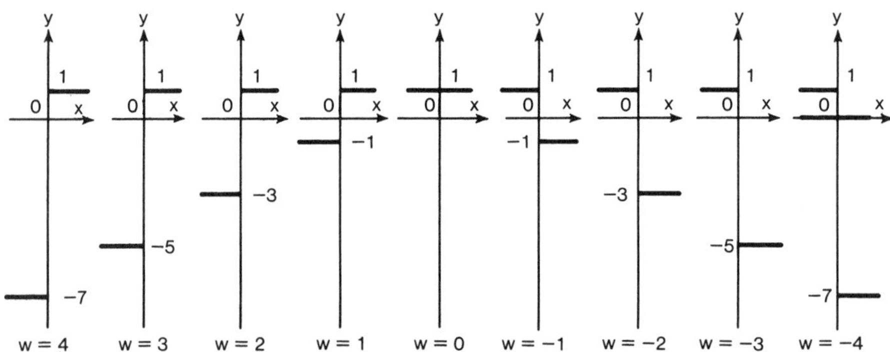

Figure 7.3 Graphs of $\text{step}_w x$.

suggesting circuits having properties not generally associated with rectification. The operation rec_w does not require amplification if $-1 \leqslant w \leqslant 1$; but amplification is required if $|w| > 1$.

In general, all-wave rectification has no unique inverse operation. However, if w and $\text{sgn}\, u$ are given, and if $w \neq \pm\frac{1}{2}$, then $\text{rec}_w u$ is invertible. This is shown in the relationship, obtained from Eq. (7.14),

$$u = \frac{\text{rec}_w u}{1 - |w| + w \,\text{sgn}\, u} \tag{7.16}$$

(It is apparent from this expression that u can "blow up" when $w = \pm\frac{1}{2}$, but has a unique solution otherwise for all u-values.) As a result of this relationship, a possible application for all-wave rectification is used as a method of *analog signal encoding* to provide secure voice communications.

The three signals $\text{rec}_w u$, w, and $\text{sgn}\, u$ can be impressed on three different carriers. For further security, the frequencies of these carriers can be shifted in response to other codes. All processing is analog. No digital coding or processing is required, although sgn u has a "digital flavor" in that it is either -1 or $+1$. If $w = 0$, then transmissions will be "in the clear," i.e., not encoded.

7.1.3 Clipping at a lower level of zero

Half-wave rectification constitutes *clipping* at a lower level of zero. This is written descriptively as

$$\text{clip}_0 u = \text{rec}_{1/2} u \tag{7.17}$$

Clipping has also been called *bounding, limiting, slicing, amplitude selection, saturation, clamping,* and *squaring*. "Saturation" is more often used to refer to a mechanism that produces approximate clipping. For example, the regions at the two ends of the linear region of an otherwise linear device are its saturation regions. "Clamping" has traditionally been used more often to mean a shift in d-c level referenced to some characteristic of a signal (usually its peak value).[1]

"Squaring" has occasionally been used in the sense of "squaring-off" (clipping) pulses in a pulse train; it is currently used almost exclusively in its mathematical sense as the multiplication of a quantity by itself.

7.2 Clipping at an Arbitrary Lower Level

To clip at an arbitrary lower level, a, we write

$$\rightarrow\rightarrow\rightarrow \quad \text{clip}_a u \equiv a + \text{rec}_{1/2}(u - a) \tag{7.18}$$

This operation is graphed in Fig. 7.4. We can also write

$$\text{clip}_a u = \tfrac{1}{2}(u + a) + \tfrac{1}{2}|u - a| \tag{7.19}$$

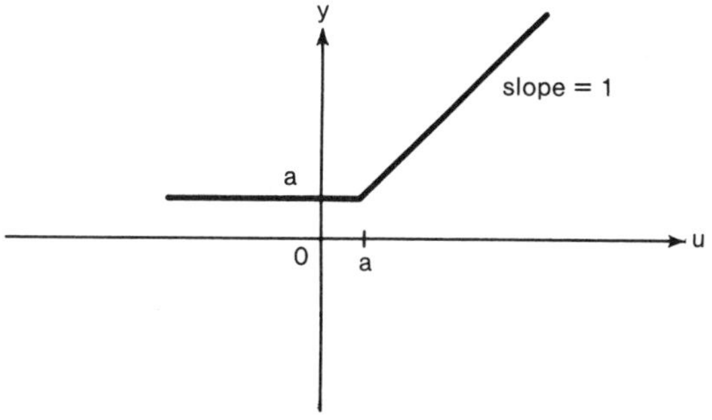

Figure 7.4 The clipping operation $\text{clip}_a u$.

[1] See for example the discussion on d-c restoration in Chapter 11 of B. Grob, *Basic Television*, 3rd ed. (New York: McGraw-Hill, 1964).

From Eq. (7.19) it is apparent that $\text{clip}_a u$ is symmetrical in the argument u and the index a so that

$$\text{clip}_a u = \text{clip}_u a \tag{7.20}$$

It can also be shown that

$$\text{clip}_a(\text{clip}_a u) = \text{clip}_a u \tag{7.21}$$

$$\text{clip}_u(\text{clip}_a u) = \text{clip}_a u \tag{7.22}$$

The derivative of $\text{clip}_a u$ (a = constant) is

$$\frac{d}{dx}\,\text{clip}_a u = S(u - a)\frac{du}{dx} \tag{7.23}$$

and of $\text{clip}_v u$ is

$$\frac{d}{dx}\,\text{clip}_v u = S(u - v)\frac{du}{dx} + S(v - u)\frac{dv}{dx} \tag{7.24}$$

7.3 Clipping at an Arbitrary Upper Level

Negative half-wave rectification constitutes *clipping at an upper level of zero*. We write

$$\text{clip}^0 u = \text{rec}_{-1/2} u \tag{7.25}$$

For clipping at an arbitrary upper level, b, we write

$$\rightarrow\rightarrow\rightarrow \quad \text{clip}^b u \equiv b + \text{rec}_{-1/2}(u - b) \tag{7.26}$$

We can also write

$$\text{clip}^b u = \tfrac{1}{2}(u + b) - \tfrac{1}{2}\,|u - b| \tag{7.27}$$

The derivative of $\text{clip}^b u$ (b = constant) is

$$\frac{d}{dx}\,\text{clip}^b u = S(-u + b)\frac{du}{dx} \tag{7.28}$$

The derivative of $\text{clip}^v u$ is

$$\frac{d}{dx}\,\text{clip}^v u = S(v - u)\frac{du}{dx} + S(u - v)\frac{dv}{dx} \tag{7.29}$$

From Eq. (7.27) it is apparent that $\text{clip}^b u$ is symmetrical in the argument u and the index b so that

$$\text{clip}^b u = \text{clip}^u \text{b} \tag{7.30}$$

Clipping at an upper level can be written in terms of clipping at a lower level as

$$\text{clip}^b u = u + b - \text{clip}_b u \tag{7.31}$$

because

$$\text{clip}_b u + \text{clip}^b u = u + b \tag{7.32}$$

which can easily be shown graphically, or by substituting the equivalent expressions from Eqs. (7.19) and (7.27) for $\text{clip}_b u$ and $\text{clip}^b u$.

The operation $\text{clip}^b u$ is graphed in Fig. 7.5.

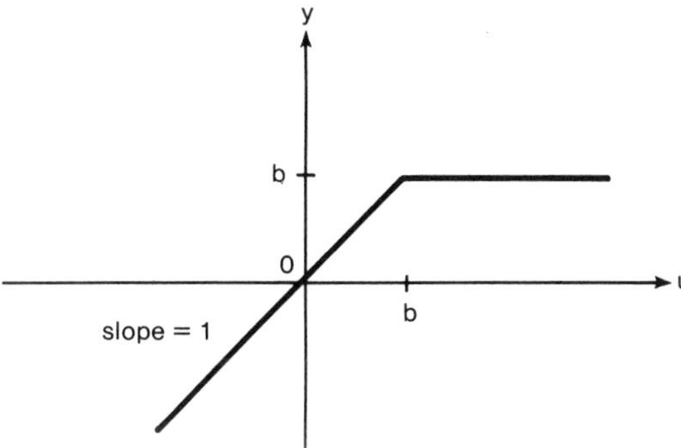

Figure 7.5 The clipping operation $\text{clip}^b u$.

7.4 Clipping at Two Levels

Clipping at two levels preserves that part of the function u lying between the lower level a and the upper level b. It is written symbolically as $\text{clip}_a^b u$. Its definition is

$$\rightarrow\rightarrow\rightarrow \quad \text{clip}_a^b u \equiv \text{clip}_a u + \text{clip}^b u - u \tag{7.33}$$

The operation $\text{clip}_a^b u$ is graphed in Fig. 7.6.

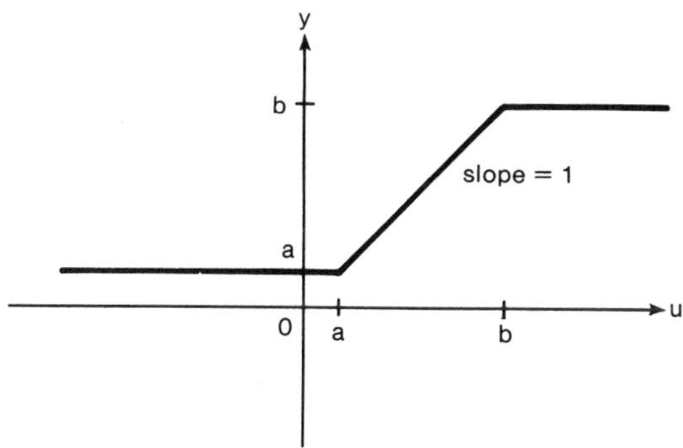

Figure 7.6 The clipping operation $\text{clip}_a^b\, u$.

There are a multitude of ways to write $\text{clip}_a^b u$. Some of these are given next:

$$\text{clip}_a^b u = a + \text{rec}_{1/2}(u - a) - \text{rec}_{1/2}(u - b) \tag{7.34}$$

$$= b + \text{rec}_{-1/2}(u - b) - \text{rec}_{-1/2}(u - a) \tag{7.35}$$

$$= a + b - u + \text{rec}_{1/2}(u - a) + \text{rec}_{-1/2}(u - b) \tag{7.36}$$

$$= u - \text{rec}_{-1/2}(u - a) - \text{rec}_{1/2}(u - b) \tag{7.37}$$

$$= \tfrac{1}{2}(a + b + |u - a| - |u - b|) \tag{7.38}$$

$$= \text{clip}^b(\text{clip}_a u), \quad a < b \tag{7.39}$$

$$= \text{clip}_a(\text{clip}^b u), \quad a < b \tag{7.40}$$

$$= \text{clip}_a u - (\text{clip}_b u - b) \tag{7.41}$$

$$= \text{clip}^b u - (\text{clip}^a u - a) \tag{7.42}$$

It is easy to show that

$$\text{clip}_a^\infty u = \text{clip}_a u \tag{7.43}$$

$$\text{clip}_{-\infty}^b u = \text{clip}^b u \tag{7.44}$$

$$\text{clip}_{-\infty}^\infty u = u \tag{7.45}$$

$$\text{clip}_a^a u = a \tag{7.46}$$

$$\text{clip}_a^b(-u) = -\text{clip}_{-b}^{-a} u \tag{7.47}$$

$$\text{clip}_b^a u = a + b - \text{clip}_a^b u \tag{7.48}$$

The derivative of $\text{clip}_a^b u$ (a, b = constant) is

$$\frac{d}{dx} \text{clip}_a^b u = \text{puls}_a^b u \frac{du}{dx} \tag{7.49}$$

If we agree to write $\text{puls}_a u = S(u - a)$ and $\text{puls}^b u = S(-u + b)$, then we also have

$$\frac{d}{dx} \text{clip}_a u = \text{puls}_a u \frac{du}{dx} \tag{7.50}$$

$$\frac{d}{dx} \text{clip}^b u = \text{puls}^b u \frac{du}{dx} \tag{7.51}$$

The derivative of $\text{clip}_u^v x$ can readily be found by noting that Theorem 4.1 applies. We write

$$\text{clip}_u^v x = u \,\text{puls}^u x + x \,\text{puls}_u x - v \,\text{puls}^v x - x \,\text{puls}_v x + v \tag{7.52}$$

and take the intrinsic derivative of each term to obtain the complete derivative of $\text{clip}_u^v x$ as

$$\frac{d}{dx}\,\text{clip}_u^v x = \text{puls}^u x \frac{du}{dx} + (\text{puls}_u x - \text{puls}_v x) + (1 - \text{puls}^v x)\frac{dv}{dx} \tag{7.53}$$

While rectification and clipping operations have traditionally been performed using diodes, they can also be performed using analog switches instead. Such a half-wave rectifier circuit is shown in Fig. 7.7. In the figure, α is a normally open analog switch and COMP is a comparator. Full- and all-wave rectification can also be performed in this manner as shown in Figs. 7.8 and 7.9, respectively.

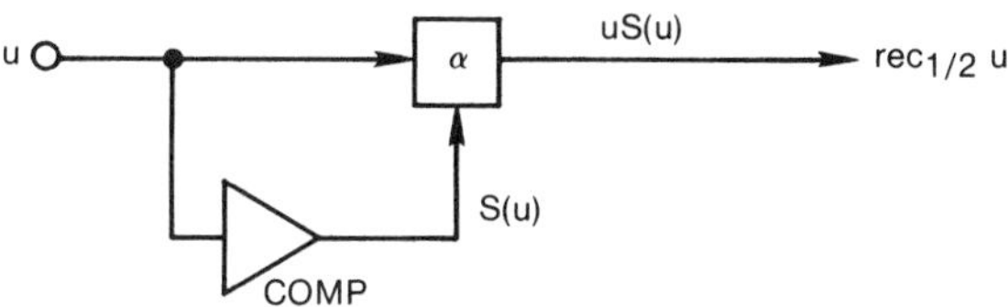

Figure 7.7 A half-wave rectifier circuit.

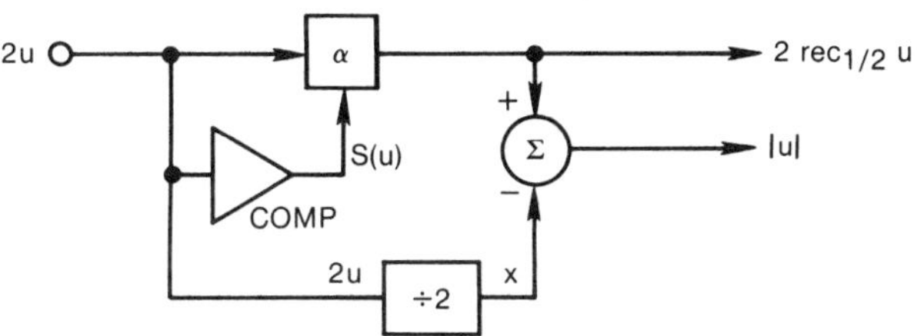

Figure 7.8 A full-wave rectifier circuit.

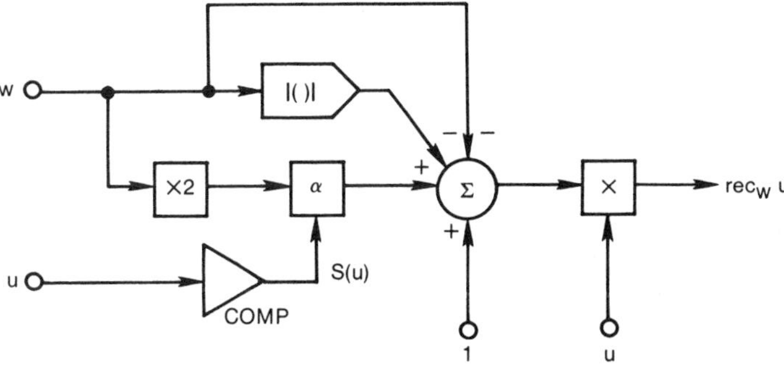

Figure 7.9 An all-wave rectifier circuit.

7.5 Performing Infinite Clipping

In the physical applications, zero-crossing detection (and hence the signum operation) can be approximated by a process known as *infinite clipping*. (Zero-crossing detection is discussed in Chapter 6.) This consists of applying the clipping operation, amplifying, clipping, amplifying, clipping, etc. (always ending with the clipping operation), until the desired accuracy is obtained.

We can write this as

$$\operatorname{sgn} u \approx \operatorname{clip}_{-1}^{1}(u/h) \tag{7.54}$$

where h is a positive constant with $h << 1$.

This method is used for the case $u = \sin x$ to produce approximate squarewaves from sinewaves.

7.6 Center Clipping

7.6.1 Definition

The *center-clipping* operation retains those portions of a clipped function normally discarded and discards those portions normally retained. We define center clipping, *ccl*, as

$$ccl_a^b u = u - \operatorname{clip}_a^b u \tag{7.55}$$

The operation $ccl_a^b u$ is graphed in Fig. 7.10. The function $ccl_{-1/2}^{1/2}(\cos x)$ is graphed in Fig. 7.11. The center clipping operation is also called a *dead zone, dead space, inert zone,* and *threshold* operation.

It is easy to show that

$$ccl_a^a u = u - a \tag{7.56}$$

$$ccl_{-\infty}^{\infty} u = 0 \tag{7.57}$$

One application for center clipping is to a problem in mechanics, discussed next.

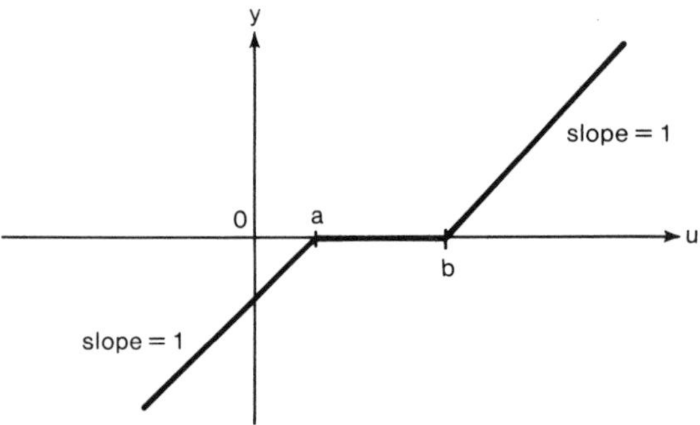

Figure 7.10 The center clipping operation.

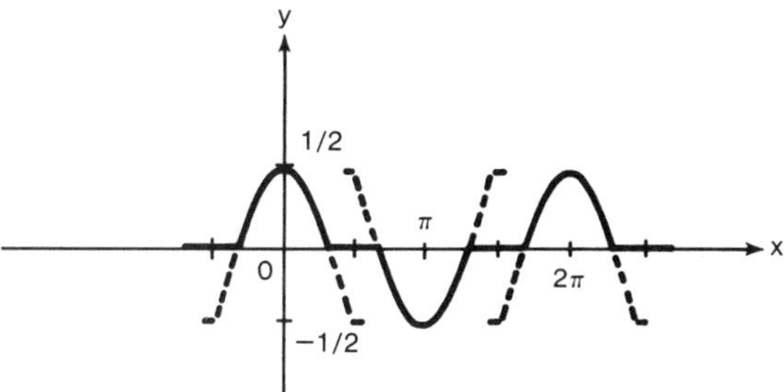

Figure 7.11 A center clipped cosinewave.

7.6.2 An application: Elementary mechanics revisited

A classic problem in elementary mechanics is the motion of a body with mass m under the influence of frictional and external forces. The situation is illustrated in Fig. 7.12.

In Fig. 7.12, $N = mg$ is the normal-force reaction to the weight mg of the body. The positive directions of applied force F, velocity v, and acceleration a are towards the right. The positive direction of the frictional force f is towards the left.

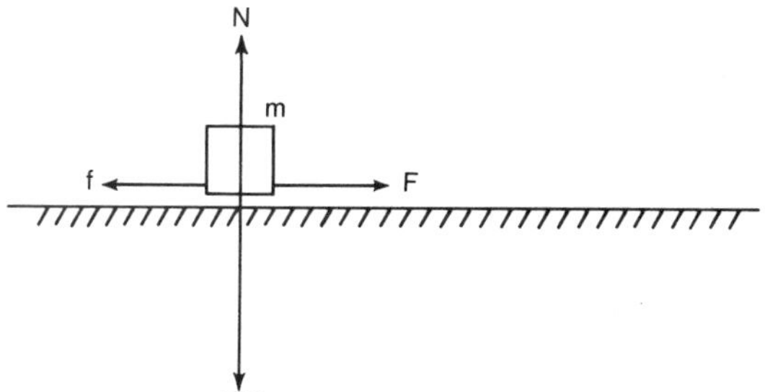

Figure 7.12 A mass under the influence of frictional and external forces.

It is pointed out in elementary physics that the static friction (i.e., the friction with $v = 0$) has a maximum magnitude f_s that is generally greater than the kinetic friction. Further, the magnitude of static friction is equal to the magnitude of the applied force up to a maximum of f_s as long as $v = 0$. The magnitude of kinetic friction f_k is essentially independent of speed. This is the frictional force acting when $v \neq 0$.

Specifically,

$$\left.\begin{array}{ll} f = f_k & \text{when } v > 0 \\ f = -f_k & \text{when } v < 0 \end{array}\right\} \tag{7.58}$$

Also,

$$\left.\begin{array}{ll} F - f = 0 & \text{when } v = 0 \text{ and } |F| \leq f_s \\ F - f = F - f_s & \text{when } v = 0 \text{ and } |F| > f_s \end{array}\right\} \tag{7.59}$$

We note that static friction always opposes the applied force; but kinetic friction opposes instead the body's velocity vector, and so may either oppose or aid the applied force. In general, it is apparent that

$$F - f = ma \tag{7.60}$$

In a typical case, we start with F, v, and a all zero. We then gradually increase F. The quantity ma under these conditions is given by the curve in Fig. 7.13. The body does not move as long as $F \leqslant f_s$. As soon as $F > f_s$, the body moves; the friction f suddenly is reduced from f_s to f_k at that instant.

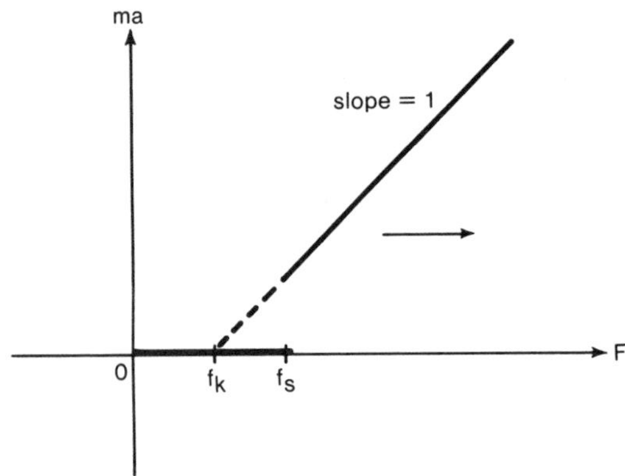

Figure 7.13 Curve of ma vs. F for the body of Fig. 7.12. A typical case.

A more general case occurs when v and a are not zero simultaneously. For example, the body may—while accelerating—pass through zero velocity in which case there is a delta pulse in the frictional force as it changes abruptly from f_k to static friction at the instant when $v = 0$. The three curves of Fig. 7.14 apply to the most general case. Any particular case can be examined by extracting the applicable portions of these three curves. Referring to Fig. 7.14, the curve for $v = 0$ is

$$ma = ccl_{-f_s}^{f_s}F, \quad v = 0; \tag{7.61a}$$

and for $v \neq 0$,

$$ma = F - f_k \operatorname{sgn} v, \quad v \neq 0 \tag{7.61b}$$

Equations (7.61a) and (7.61b) can be combined as

$$ma = \delta^0(v)\, ccl_{-f_s}^{f_s}F + \delta^{0\prime}(v)(F - f_k \operatorname{sgn} v) \tag{7.62}$$

where we have written $\delta^{0\prime}(v)$ for $1 - \delta^0(v)$, the one's complement of $\delta^0(v)$. Equation (7.62) is good for all values of v and F.

Equation (7.62) shows vividly how the acceleration of the body depends on the applied force and on the velocity of the body. Equation

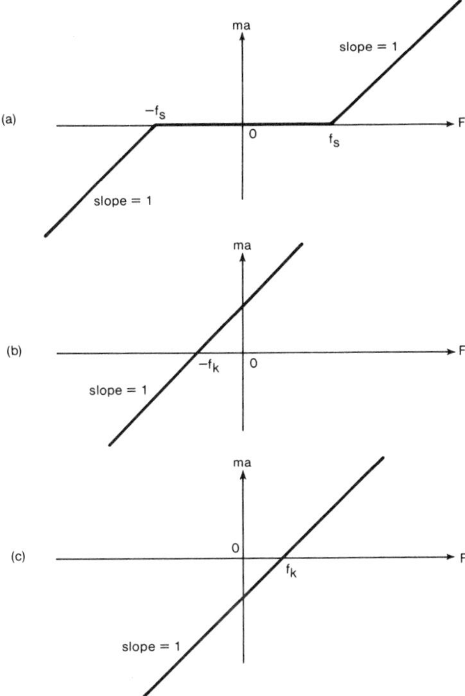

Figure 7.14. Curve of *ma* vs. *F* for the body of Fig. 7.12. A general case. (a) $v = 0$; (b) $v > 0$; (c) $v < 0$.

(7.62) further illustrates the first physical interpretation we have encountered for a vestigial delta pulse, as the ideal frictional force of an accelerating body acted on by a constant force $F > f_s$ as it passes through zero velocity. The graph of f as a function of v is shown in Fig. 7.15. Its equation is

$$f = \delta^0(v)f_s + \delta^{0\prime}(v)f_k \operatorname{sgn} v, \quad F > f_s \tag{7.63}$$

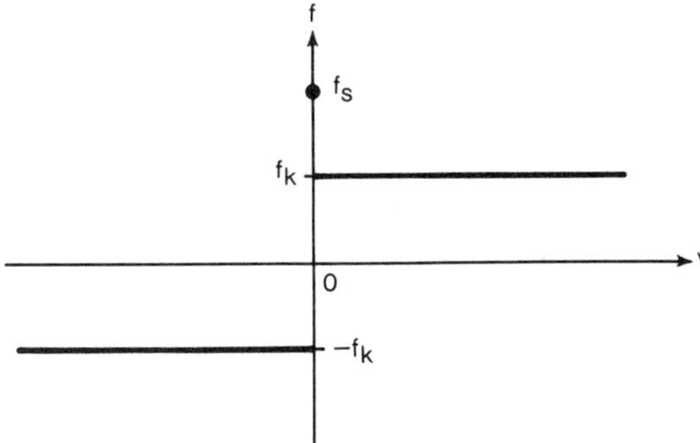

Figure 7.15 Frictional force, f, as a function of velocity, v, for the body of Fig. 7.12.

We note that the point at $(v, f) = (0, f_s)$ is not the content of the delta pulse; it is the delta pulse itself.

Going back to Fig. 7.13 for a moment, if F is reduced (with the body moving) to less than f_s, we see that the original curve is not retraced. Instead hysteresis exists as shown in Fig. 7.16. Functions with hysteresis form the subject matter of the next chapter.

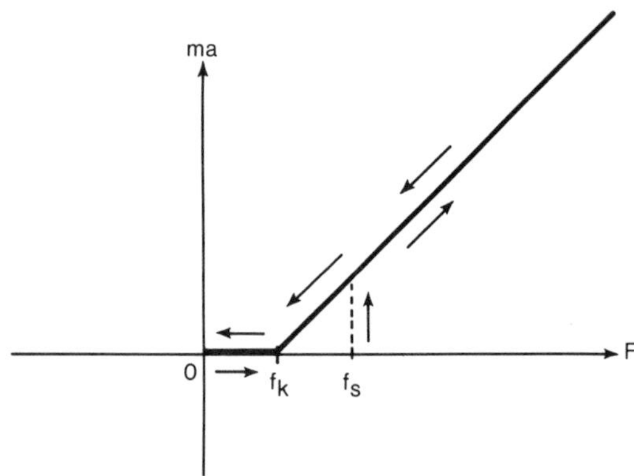

Figure 7.16 Curve showing hysteresis in a typical case.

7.6.3 Pulse functions synthesized from center-clipped functions

Previous chapters have discussed ways of synthesizing pulse functions. Another way is possible using the center-clipping operation. This is discussed next.

We recall that $\delta^0(u)$ is zero when $u \neq 0$ and one when $u = 0$. Thus the wide pulse function $\text{puls}_a^b x$ results if we let u be a center clipped function. That is,

$$\text{puls}_a^b x = \delta^0(ccl_a^b x) \tag{7.64}$$

The block diagram of this equation, illustrating how this pulse-forming technique is implemented, is shown in Fig. 7.17. It will be recalled from Chapter 2 that δ^0 is an infinitely-narrow window function or *coincidence detector*. It is labeled COINC in the block diagram.

Figure 7.17 Producing a pulse function by using a coincidence detector.

7.7 Ramp Functions

In Chapter 6 it was shown that the integral of a pulse function is a ramp function. The clipping function $y = \text{clip}_a^b x$ is a *linear ramp function* with unit slope and breakpoints at $(x, y) = (a, a)$ and (b, b). Other ramp functions can be written in terms of the clipping function.

For a ramp with slope m, y-intercept y_0 (for the ramp-extended), and breakpoints at $x = a$ and b with $a < b$, we write

$$\rightarrow\rightarrow\rightarrow \quad y = m\,\text{clip}_a^b x + y_0 \tag{7.65}$$

This function is shown in Fig. 7.18. The derivative of Eq. (7.65) is

$$\frac{dy}{dx} = m\,\text{puls}_a^b x \tag{7.66}$$

If $a = -\infty$ or $b = \infty$, a *semi-infinite* linear ramp function results.

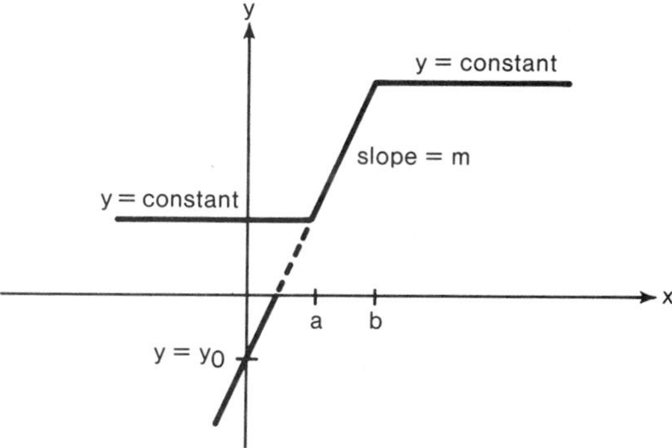

Figure 7.18 A ramp function.

7.8 Diode Characteristics

The (hypothetical) *ideal switching diode* has zero resistance for one direction of an applied voltage and zero conductance for the other direction. The series combination of a resistor with such a device has the step-function conductance characteristic shown in Fig. 7.19. This characteristic has the equation

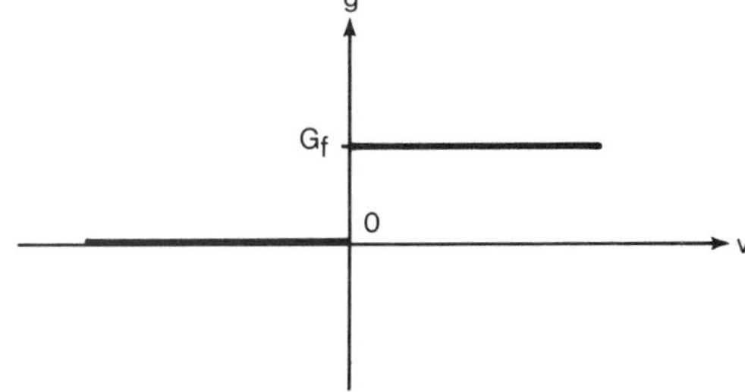

Figure 7.19 Conductance, g, vs. voltage, v, characteristic of an ideal diode and resistor in series.

$$g = G_f S(v) \tag{7.67}$$

where g is the network conductance, G_f is the resistor conductance, and v is the applied voltage. The current, y, through this combination is

$$y = gv = G_f v S(v) \tag{7.68}$$

The quantity $vS(v)$ is the half-wave rectification of v.

If we now parallel the diode with a second resistor (conductance = G_2) as shown in Fig. 7.20, we obtain the resultant conductance characteristic shown in Fig. 7.21 with equation

$$g = G_f S(v) + G_r S(-v) \tag{7.69}$$

where

$$G_r = G_f \| G_2 \tag{7.70}$$

(The *parallel operation,* $\|$, is discussed in Appendix E.) The current through this combination is

$$y = G_f v S(v) + G_r v S(-v) \tag{7.71}$$

We recognize $vS(-v)$ as the negative half-wave rectification of v.

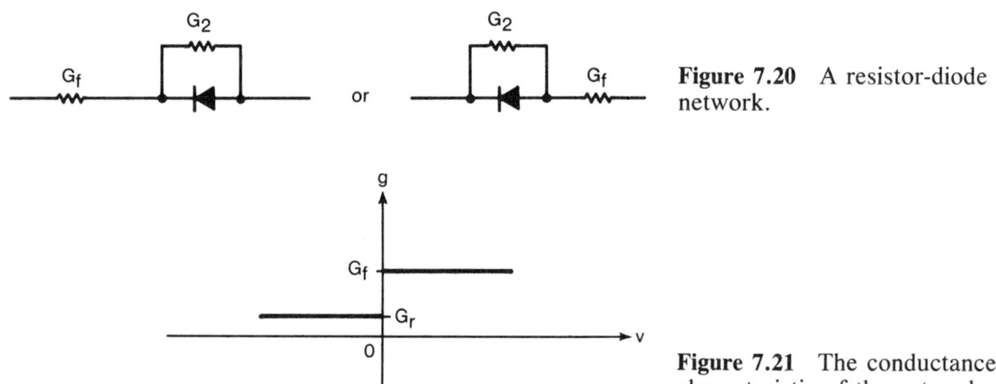

Figure 7.20 A resistor-diode network.

Figure 7.21 The conductance characteristic of the network of Fig. 7.20.

A real switching diode has a macroscopic (large scale) characteristic that more closely resembles Eq. (7.71) than that of a switch. In the circuit just described, G_f and G_r simulate the *forward* and *reverse* conductances, respectively, of a real diode.

It is apparent from the above equations that switching diodes are well suited for use in rectifying, clipping, and center clipping circuits. Indeed, they find widespread use in such circuits.

7.9 Diode Function Generation

A method of function synthesis called "function splicing" that uses sequential gating is described in Chapter 6. Another method called *diode function generation*[2] is described below. This method permits synthesis of arbitrary piecewise-linear functions and operations.

As an example, the function shown in Fig. 7.22a can be decomposed into the two semi-infinite ramps shown in Figs. 7.22b and 7.22c, so that

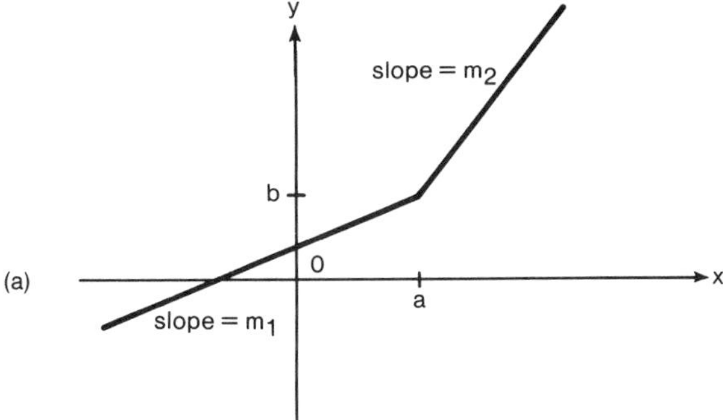

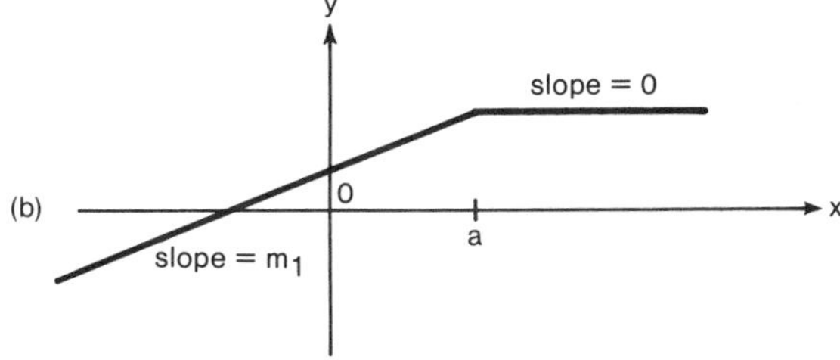

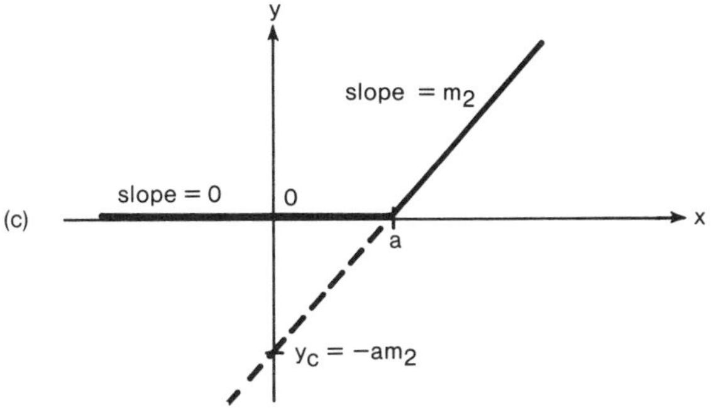

Figure 7.22 Showing how a compound ramp (a) can be decomposed into two simple ramps (b) and (c).

[2]So-called because it has traditionally been performed using diodes.

the function is the sum of these two ramps. Therefore, the equation of the function of Fig. 7.22a is readily found to be

$$y = m_1 \text{clip}^a x + m_2 \text{clip}_a x + b - a(m_1 + m_2) \tag{7.72}$$

Piecewise-linear functions having more than two segments can be similarly synthesized as a sum of ramp functions.

For three segments we have

$$\begin{aligned} y = {} & m_1 \text{clip}^{a_1} x + m_2 \text{clip}_{a_1}^{a_2} x + m_3 \text{clip}_{a_2} x \\ & + b - a_1(m_1 + m_2) - a_2 m_3 \end{aligned} \tag{7.73}$$

Refer to Fig. 7.23. From this, the pattern for any number of segments is clear. For four segments we have

$$\begin{aligned} y = {} & m_1 \text{clip}^{a_1} x + m_2 \text{clip}_{a_1}^{a_2} x + m_3 \text{clip}_{a_2}^{a_3} x \\ & + m_4 \text{clip}_{a_3} x + b - a_1(m_1 + m_2) - a_2 m_3 - a_3 m_4 \end{aligned} \tag{7.74}$$

By treating the above relationships as operations instead of functions (i.e., by substituting u for x), many more kinds of functions can be synthesized.

7.10 A Universal Clipping Module or Integrated Circuit

One can envision a *universal clipping module* or integrated circuit that can be wired externally to perform as a single- or two-level clipper, center clipper, zero-crossing detector, half-wave rectifier, negative half-wave rectifier, full-wave rectifier, ramp function generator, etc. It might contain input and output amplifiers with gain to be set by external resistors or voltages.

The universal clipping module is shown in block diagram form in Fig. 7.24. The manner in which it would be connected to perform various functions is shown in Table 7.1 (see page 124).

7.11 Overview

This and the previous chapter have presented a substantial number of operations of interest to the analog circuit designer. In the next chapter, it is shown how some of these operations can be used in the description of hysteretic processes.

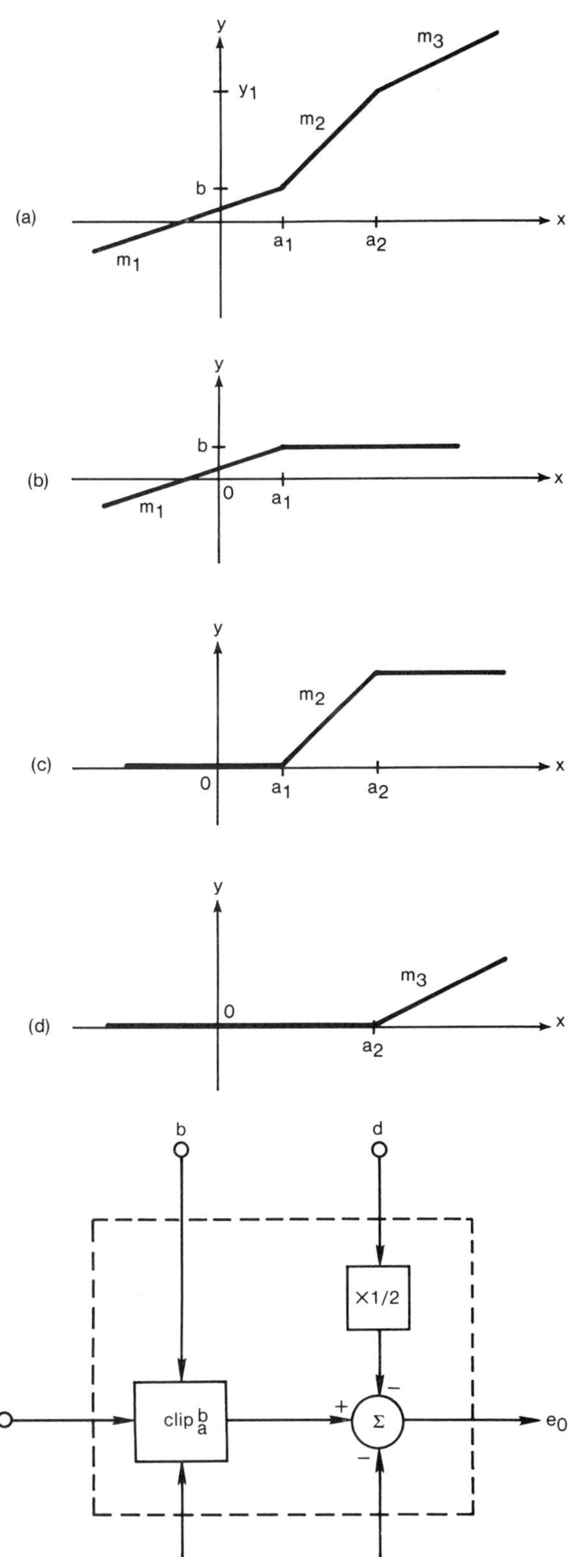

Figure 7.23 Showing how another compound ramp (a) can be decomposed into three simple ramps (b), (c), and (d).

Figure 7.24 A universal clipping module.

TABLE 7.1 Connections for the Universal Clipping Module

OPERATION	EXPRESSION FOR e_0	a TERMINAL	b TERMINAL	JUMPERS	BASIC FUNCTION
Unity-gain buffer	e_i	open	open	none	$\text{clip}_{-\infty}^{\infty} e_i$
Clipper	$\text{clip}_a^b e_i$	E_a	E_b	none	$\text{clip}_a^b e_i$
Center clipper	$-\text{ccl}_a^b e_i$	E_a	E_b	connect "c" to input	$\text{clip}_a^b e_i - e_i$
Half-wave rectifier	$\text{rec}_{1/2} e_i$	ground	open	none	$\text{clip}_0^{\infty} e_i$
Negative half-wave rectifier	$\text{rec}_{-1/2} e_i$	open	ground	none	$\text{clip}_{-\infty}^{0} e_i$
Full-wave rectifier	$\frac{1}{2}\lvert e_i\rvert$	ground	open	connect "d" to input	$\text{clip}_0^{\infty} e_i - \frac{1}{2} e_i$
Negative full-wave rectifier	$-\frac{1}{2}\lvert e_i\rvert$	open	ground	connect "d" to input	$\text{clip}_{-\infty}^{0} e_i - \frac{1}{2} e_i$
Zero-crossing detector	$\approx 0.1 \text{ sgn } e_i$	$e_{-0.1}$	$e_{0.1}$	none	$\text{clip}_{-0.1}^{0.1} e_i$

8

HYSTERETIC FUNCTIONS AND PROCESSES

8.1 What Is Hysteresis?

In this chapter, methods are developed for expressing hysteretic processes in terms of relaxation operations. Then, with analytic expressions available, it becomes a simple matter to produce block diagrams that detail these multiple-valued processes.

Infinitesimal hysteresis occurs naturally in functions such as sgn x, but we must deliberately introduce hysteresis if we want it to affect more than a single point.

The term "hysteresis" perhaps evokes a mental picture of a "hysteresis loop"—the BH curve of magnetism. It may evoke a picture of backlash in a gear train. Or it may evoke something as common as the toggling action of a light switch. Hysteresis is all of these things and more.

Hysteresis has been defined as the influence of the past history of a system upon its present behavior. For the present volume we use a definition somewhat less general. For systems containing a single input and a single output, nonhysteretic and hysteretic systems can be contrastingly defined as follows: A nonhysteretic system is one whose transfer characteristic is not affected by the input, and a hysteretic system is one whose transfer characteristic displays a *repeatable dependence* on the input, "repeatable" meaning that adaptive systems and catastrophic systems are ruled out.

In this chapter we are concerned with methods of expressing the multiple transfer characteristics of hysteretic systems as compact functions.

8.2 Categories of Hysteresis

In the applications we encounter two general categories of hysteresis that we shall call *level triggered* and *rate triggered.* That is, either a change in level or a change in the rate-of-change of the input (or output) triggers a change in the transfer characteristic. The change in transfer characteristic can be either a *discrete change*[1] between two (possibly more) predetermined functions, as in toggling; or it can be a gradual transition to one of a continuum of functions, as in magnetic phenomena.

The various transfer characteristics of a given hysteretic relaxation function are called *branches.* The change between transfer characteristics is called *branch transfer.*

8.3 Functions That Display Level-Triggered Hysteresis

Among its many other uses, the Heaviside step function can serve as a *unit hystor,* the term applied to a quantity that selects among alternate branches. In this respect, it resembles an operator in operational mathematics. However, the unit hystor is used as an ordinary multiplicative quantity.

For example, $S(u)$ can be made to be either 0 or 1, as we wish, depending on the makeup of u. For a level-triggered hysteresis, we want $S(u)$ to change from 0 to 1 (or vice-versa) at a given level. A particular case of level-triggered hysteresis is diagrammed in Fig. 8.1.

An expression for the hysteretic function in Fig. 8.1 is

$$y = S(y)\text{sgn}(x + a) + S(-y)\text{sgn}(x - a) \tag{8.1}$$

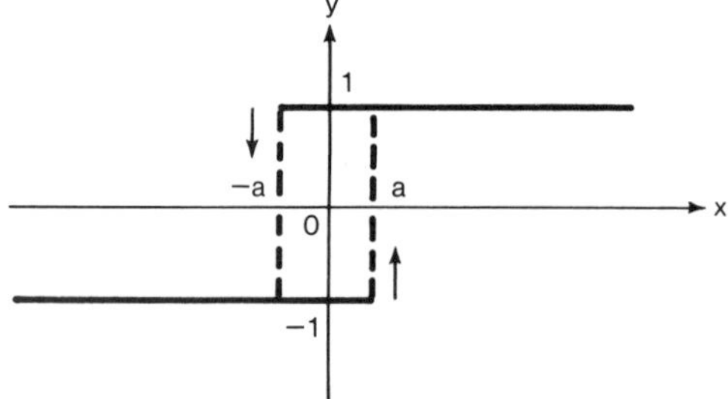

Figure 8.1 Level-triggered hysteresis.

[1] An abrupt step change if along a line input = constant, a dead-zone type change if along a line output = constant.

In the system of Eq. (8.1) and Fig. 8.1, y can have one of two values: $+1$ and -1. When $y = 1$, $S(y) = 1$ and $S(-y) = 0$; when $y = -1$, $S(y) = 0$ and $S(-y) = 1$. Thus one or the other of the branches $\operatorname{sgn}(x + a)$ and $\operatorname{sgn}(x - a)$ is selected at any given time. This is the desired action. Equation (8.1) is an *implicit relationship* of x and y; that is, it is not in the form $y = f(x)$ or $x = g(y)$. Branch transfer occurs along the lines $x = -a$ and $x = a$ as required.

Another way to generate the function of Fig. 8.1 is to write simply

$$y = \operatorname{sgn} u \tag{8.2}$$

where u is a function such as the one shown in Fig. 8.2. For example, let

$$u = ccl{-}{}^{a}_{a}x \tag{8.3}$$

Since *sgn* exhibits infinitesimal hysteresis, its value at $u = 0$ depends on whether u was previously positive or negative. Thus, the infinitesimal hysteresis of the signum function is magnified by its argument to a finite hysteresis. The function u is called a *magnifying function.*

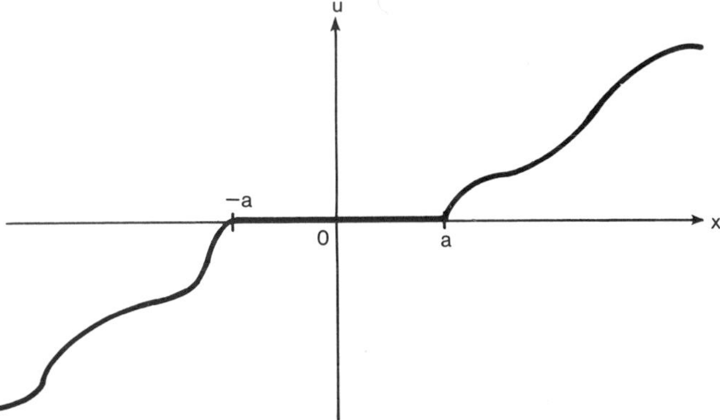

Figure 8.2 A magnifying function.

Another example of level-triggered hysteresis is now given. The current (y) vs. voltage (v) characteristic of a neon lamp is nearly represented by the hysteretic curve shown in Fig. 8.3. The voltages a and b represent the extinction and firing voltages of the neon lamp, respectively. When the neon lamp fires (neon ionizes), it has a high conductance, G; and when the lamp extinguishes (neon de-ionizes), it has a low conductance, nominally zero. Using the method of magnifying functions, the curve of Fig. 8.3 can be represented by the expression

$$y = G[S(ccl^{-a}_{-b}v) + S(ccl^{b}_{a}v)](v - c) \tag{8.4}$$

where c is the v-axis intercept of the sloping locus extended.

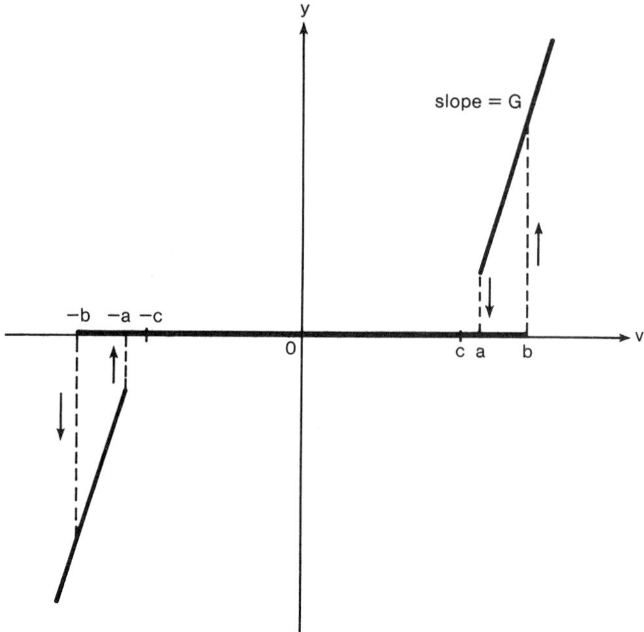

Figure 8.3 Current (y) vs. voltage (v) characteristic of a neon lamp.

A neon lamp can serve as the hysteretic element in oscillators that produce approximate sawtooth waves. A circuit diagram of such an oscillator is shown in Fig. 8.4. This is a classic circuit. The constant-current source can be replaced by a resistor if less accuracy can be tolerated. If a unijunction transistor is used in place of the neon lamp, then narrow pulses are also produced. Otherwise the operation is essentially the same as with the neon lamp. This latter circuit is diagrammed in Fig. 8.5.

The method of magnifying functions can be generalized to apply to the general hysteretic function of Fig. 8.6-a as follows. First, we shrink the band of hysteresis to zero as shown in Fig. 8.6-b and write for this curve the expression

$$y = f(x)S(x - c) \tag{8.5}$$

Since $ccl_c^c x = x - c$, Eq. (8.5) can also be written

$$y = f(x)S(ccl_c^c x) \tag{8.6}$$

It is now a simple matter to write the expression for the curve of Fig. 8.6-a as

$$y = f(x)S(ccl_a^b x) \tag{8.7}$$

The block diagram of this equation appears in Fig. 8.7.

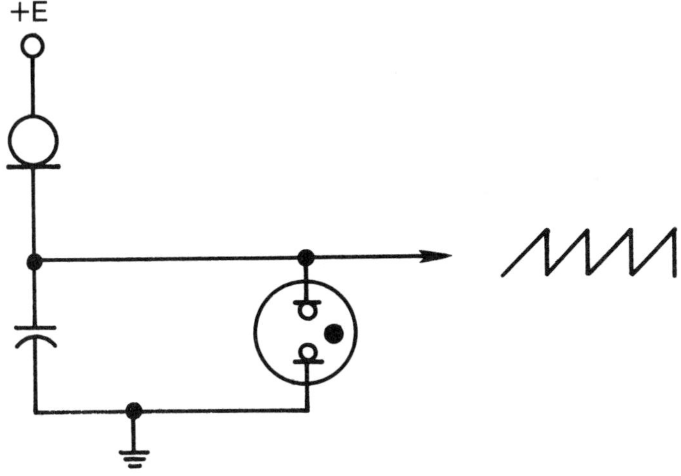

Figure 8.4 A sawtooth oscillator using a neon lamp.

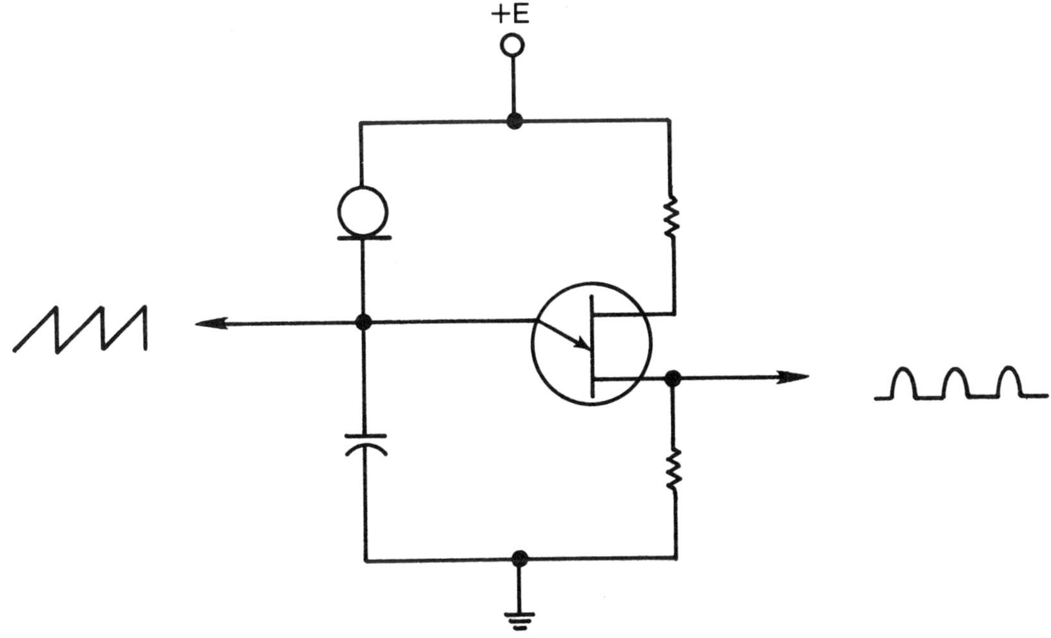

Figure 8.5 A sawtooth oscillator using a unijunction transistor.

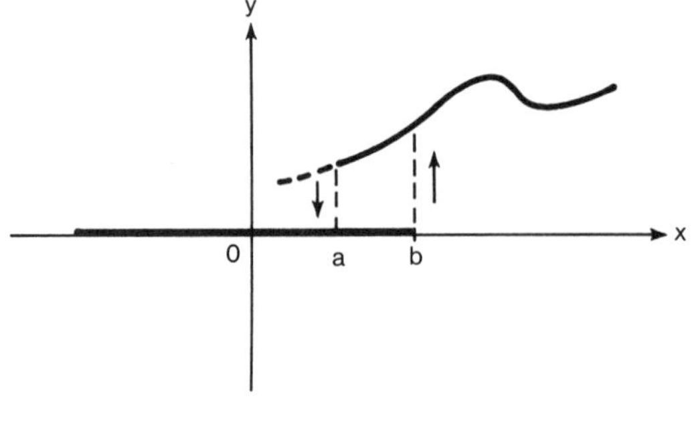

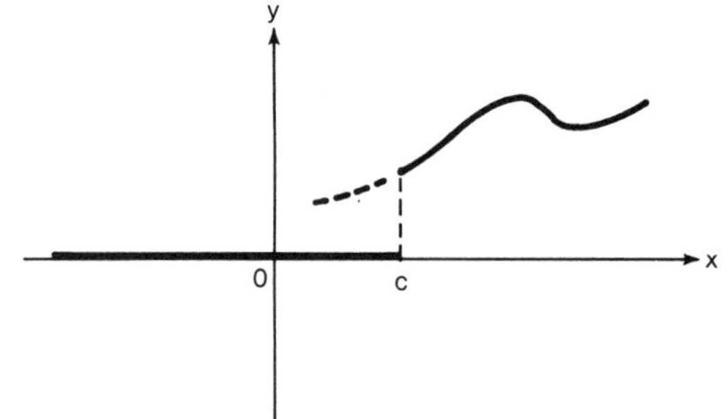

Figure 8.6 (a) A general hysteretic function; (b) the band of hysteresis shrunk to zero.

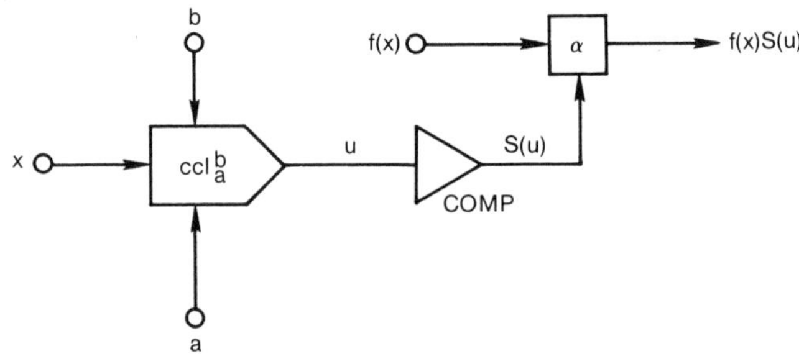

Figure 8.7 Block diagram of the function of Fig. 8.6a.

Examples of level-triggered hysteresis are toggling, oilcanning, and latchup. This is also the kind of hysteresis displayed by a two-level comparator with positive feedback (Schmitt trigger). Such a circuit is shown in Fig. 8.8. The amount of hysteresis can be controlled by varying the amount of positive feedback with the potentiometer.

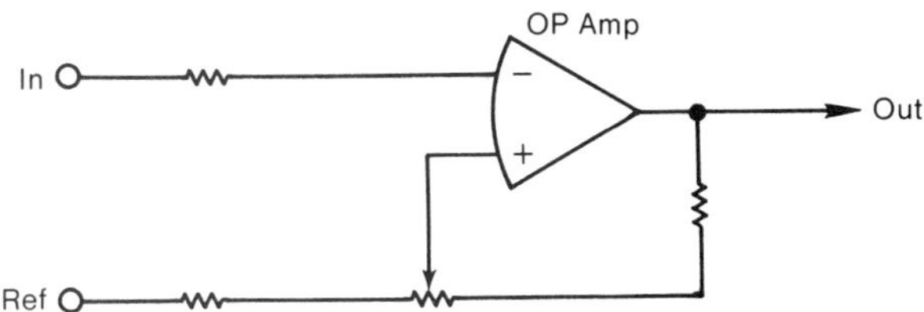

Figure 8.8 A comparator with hysteresis.

This kind of hysteresis is also characteristic of a wide variety of multivibrators including astable (free-running), monostable (one-shot), and bistable (flip-flop) varieties.

Devices that rely on hysteresis to make them oscillate are traditionally called *relaxation oscillators*.[2] We might extend this term to include all generators of periodic relaxation functions.[3]

8.4 Rate-Triggered Hysteresis

In one form of rate-triggered hysteresis, branch transfer occurs upon reversal of the input direction of progression. The unit hystors in this case take the form $S(\dot{x})$ and $S(-\dot{x})$ where[4]

$$S(\dot{x}) = \tfrac{1}{2} + \tfrac{1}{2}\frac{dx/dt}{|dx/dt|} = \tfrac{1}{2} + \tfrac{1}{2}\frac{dx}{|dx|} \tag{8.8}$$

$$S(-\dot{x}) = \tfrac{1}{2} - \tfrac{1}{2}\frac{dx/dt}{|dx/dt|} = \tfrac{1}{2} - \tfrac{1}{2}\frac{dx}{|dx|} \tag{8.9}$$

When progression is in the $+x$ direction, $\dot{x}$ is positive so that $S(\dot{x}) = 1$ and $S(-\dot{x}) = 0$; when progression is in the $-x$ direction, $\dot{x}$ is negative so that $S(\dot{x}) = 0$ and $S(-\dot{x}) = 1$.

This kind of hysteretic function can be written

$$y = S(\dot{x})f_1(x) + S(-\dot{x})f_2(x) \tag{8.10}$$

giving $y = f_1(x)$ for positive progression (+prog) and $y = f_2(x)$ for negative progression (−prog). If $f_1(x)$ and $f_2(x)$ are alike except over a finite range, then a *finite hysteresis* exists. If they differ over all x, then an *infinite hysteresis* exists. Or if they differ from $x = a$ to ∞ (or $-\infty$), then a *semi-infinite hysteresis* exists.

[2] B. Chance, *Waveforms* (New York: McGraw-Hill, 1949), p. 165.

[3] Minorsky reserves the term *relaxation oscillator* for devices that oscillate because of hysteresis, and he uses the term *impulse-excited oscillator* for devices such as the bouncing-ball oscillator of Section 4.4.2. N. Minorsky, *Introduction to Non-linear Mechanics* (Ann Arbor, Mich.: J. W. Edwards, 1947), p. 384.

[4] Here dt is taken to be alway positive.

If a reversal of progression occurs at some point, then branch transfer occurs at that point. If progression simply stops and resumes without reversing, then there is no branch transfer. This results from the fact that the unit hystors themselves display infinitesimal hysteresis.

The derivative of Eq. (8.10) is

$$\frac{dy}{dx} = S(\dot{x})\frac{d}{dx}f_1(x) + S(-\dot{x})\frac{d}{dx}f_2(x) \tag{8.11}$$

because x and $\dot{x}$ are orthogonal quantities; i.e., $\dot{x}$ does not depend on x. The integral of Eq. (8.10) is

$$\int ydx = S(\dot{x})\int f_1dx + S(-\dot{x})\int f_2(x)\,dx \tag{8.12}$$

again because x and $\dot{x}$ are orthogonal quantities.

8.5 Functions That Display Linear Hysteresis

In one kind of rate-triggered hysteresis, sometimes called *linear hysteresis,* branch transfer occurs along a line y = constant. This is diagrammed in Fig. 8.9.

For positive progression, the transfer characteristic is

$$f_1(x) = \text{clip}_b(x - a) \tag{8.13}$$

and for negative progression, the transfer characteristic is[5]

$$f_2(x) = \text{clip}^c(x + a) \tag{8.14}$$

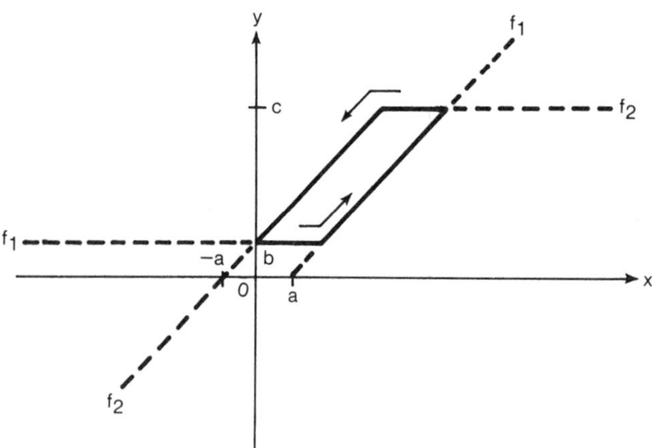

Figure 8.9 Linear hysteresis.

[5]Equations (8.13) and (8.14) constitute a *branch pair.*

Therefore the relaxation function that expresses the entire hysteretic function is

$$y = S(\dot{x})\mathrm{clip}_b(x - a) + S(-\dot{x})\mathrm{clip}^c(x + a) \tag{8.15}$$

where a is a constant, b and c are parameters with b being the y-value at which reversal from −prog to +prog occurs and c being the y-value at which reversal from +prog to −prog occurs. In this sense, linear hysteresis is *level dependent*.

An example of linear hysteresis is the backlash in a gear train. The constant $2a$ is the amount of backlash. A generalization of this kind of hysteresis is discussed in the next section.

8.6 Functions That Display Magnetic Hysteresis

The BH curve of *magnetic hysteresis* is also rate triggered and level dependent. As in the previous case, branch transfer is triggered by a reversal in the direction of progression; but unlike the previous case, the particular branch to which transfer occurs depends on the transfer point. Thus, there are virtually an infinite number of branches to which transfer can occur. We say that there is a *continuum of transfer characteristics*. Branch transfer is immediate, so that there is no step or dead-zone type change. That is, the "old" branch and the "new" branch have the *transfer point* in common.[6]

This type of hysteresis is characteristic of a system consisting of a magnetic material in the magnetic field produced by a coil carrying current, hence the name "magnetic hysteresis."

A typical branch pair is shown in Fig. 8.10. The x-value is proportional to the current in the coil, and the y-value is proportional to the resulting magnetic field. The y-axis intercept represents the magnetism remaining in the core when the coil current is zero.

If the magnetizing current varies periodically between constant maximum and minimum peaks as in Fig. 8.10, then only two branches need be considered. The two desired curves f_1 and f_2 might then be represented by doubly translated and scaled expressions of the form $\tanh x$ (which is asymptotic to $\mathrm{sgn}\, x$), or of the form $\sinh^{-1} x$ (which is asymptotic to $\mathrm{sgn}\, x \,\mathrm{Ln}\, |2x|$).[7] Equation (8.10) applies.

[6] This is also the case with linear hysteresis as defined by Eq. (8.15).

[7] Sheingold calls $\sinh^{-1}$ the *a-c log* because it is nearly logarithmic for large magnitudes of the argument, continuous through the point (0, 0), and is an odd ("bipolar") function.

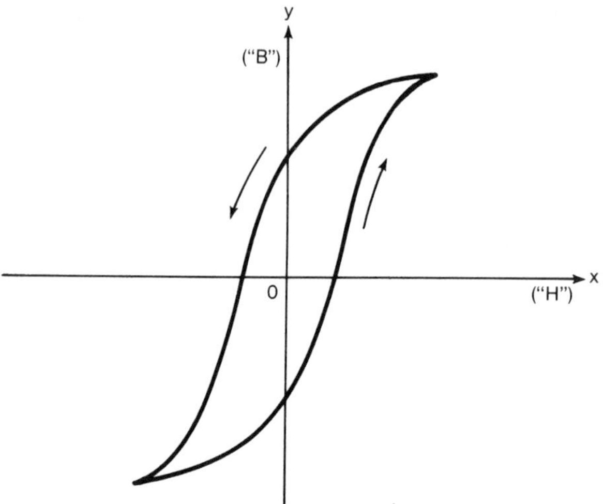

Figure 8.10 Magnetic hysteresis.

If symmetry exists (as is generally the case with magnetic materials) such that $f_2(x) = -f_1(-x)$, then simplifications result. Equation (8.10) can be replaced with

$$y = S(\dot{x})f(x) - S(-\dot{x})f(-x) \tag{8.16}$$

where f has been written for f_1. An equivalent way to write this is

$$y = \text{sgn}\,\dot{x}f(x\,\text{sgn}\,\dot{x}) \tag{8.17}$$

This latter form is diagrammed in Fig. 8.11.

The above expressions apply when x is a given periodic function of t. When x is any arbitrary function of t, the hysteretic function changes as the point of reversal changes so that there are an infinite number of possible branches for each direction of progression. Thus instead of $f(x)$,

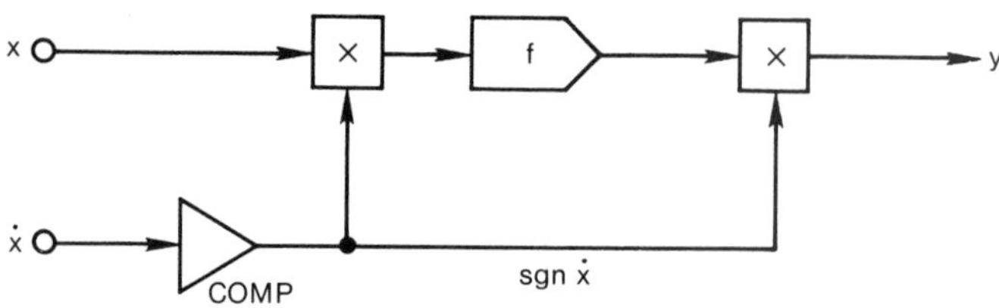

Figure 8.11 Block diagram of magnetic hysteresis.

we must use $f(x, x_i, y_i)$ where x_i and y_i are parameters that are the xy coordinates of the *initial point,* i.e., the point from which a given branch originates. That is, *the curve depends on the initial point.*

This brings us to the limit of the current state of the art. It seems that this approach to the study of magnetic hysteresis—when fully developed—could form the basis for an entire volume!

8.7 Comment

With this chapter, the study of the bulk of the traditional subjects of analog circuit design ends.[8] The remaining chapters deal with advanced concepts in relaxation analysis.

We begin, in the following chapter, with a look at infinitely-discontinuous functions and other functions with singularities. The concept of "complete derivative" is extended to infinitely-discontinuous functions. Also, it is shown how such functions can be integrated across discontinuities.

[8] Some exceptions are coordinate conversion, logarithmic amplifiers, and multifunction converters. These are discussed in Chapter 11.

9

FUNCTIONS WITH SINGULARITIES

9.1 Unique Characteristics of Functions with Singularities

Equation (1.6) categorizes functions as "continuous" or "discontinuous." Equation (1.7) further categorizes functions as "finitely discontinuous" or "other discontinuous," the latter being those for which one or both of the limits in Eq. (1.7) do not exist. The present chapter deals with functions in the "other discontinuous" category. We consider functions with singularities.

Two kinds of singularities are considered. In both kinds, the function consists of two branches that are continuous, at least in the neighborhood of the singularity at $x = a$. In the first kind, the two branches are asymptotic to the line $x = a$, one branch tending towards $+\infty$ and the other branch tending towards $-\infty$. An example of this kind of function is $1/x$. The function $1/x$ satisfies the condition that $\beta = -1$ (with $a = 0$), where

$$\beta = \lim_{x \to a} \frac{f(x)}{f(2a - x)} \tag{9.1}$$

Such a function is said to have an *infinite discontinuity* at $x = a$. The function $y = 1/x$ is graphed in Fig. 9.1.

In the second kind of function, the two branches are also asymptotic to the line $x = a$, but the two branches tend towards infinity in the same

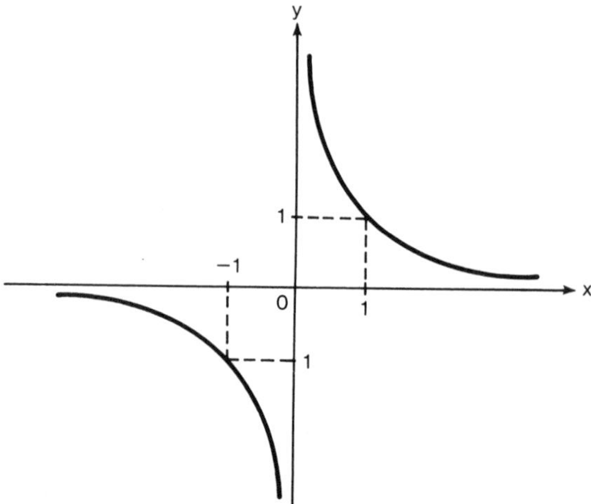

Figure 9.1 Graph of $y = 1/x$.

direction. An example of this kind of function is $1/|x|$. The function $1/|x|$ satisfies the condition that $\beta = 1$ (with $a = 0$) in Eq. (9.1). Such a function is said to have a *cusp* or *break at infinity* at $x = a$. The function $y = 1/|x|$ is graphed in Fig. 9.2.

These ideas are summarized as follows:

Lemma 9.1

Given: $F(x)$ is continuous except for an infinite discontinuity at $x = a$, i.e., $\beta = -1$.

→→→ $G(x) = \pm\ \text{sgn}(x - a)\, F(x)$ has a break at infinity at $x = a$; i.e., $\beta = 1$.

Proof of the lemma is given in Appendix A.

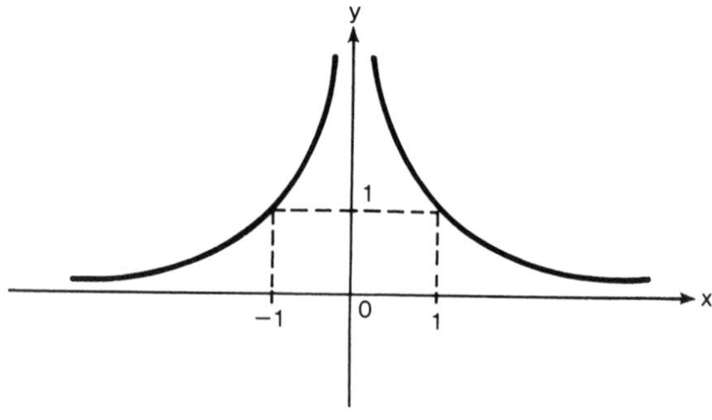

Figure 9.2 Graph of $y = 1/|x|$.

9.2 Obtaining Complete Derivatives of Infinitely Discontinuous Functions

Standard tables of derivatives of algebraic and transcendental functions—the functions of standard analysis—give intrinsic derivatives. This must be so because none of them contain delta functions. This presents no problem for functions that are continuous, because the intrinsic derivative is the complete derivative for such functions.

It also presents no problem for functions that have a break at infinity. Although such functions are categorized as discontinuous by the rules of standard analysis, *they can have no jump* because by Eq. (9.1) they are symmetric functions relative to and in the neighborhood of $x = a$. (For example, would it be a "jump up" or a "jump down"?) Therefore, they can have no delta functions in their derivatives, and the intrinsic derivative of such a function must also be its complete derivative. Thus Theorem 4.1 applies to such functions. An example follows.

Example A. $\dfrac{d}{dx}|x|^{-1} = \dfrac{d}{dx}x^{-1}\operatorname{sgn} x = -x^{-2}\operatorname{sgn} x = -|x|^{-1}x^{-1}.$

Alternatively,

$$\frac{d}{dx}|x|^{-1} = -|x|^{-2}\operatorname{sgn} x = -x^{-2}\operatorname{sgn} x$$

In the first part of this example, sgn x differentiates like a constant because its intrinsic derivative is zero.

Infinitely discontinuous functions are a different matter; such a function has an infinite jump. One example of an infinitely discontinuous function from standard analysis has already been given: $1/x$. Other examples are tangent, cotangent, secant, and cosecant functions. These functions are infinitely discontinuous and their published derivatives are not their complete derivatives.

The complete derivatives of functions with infinite discontinuities can be found by a procedure similar to that used for functions with finite discontinuities. The procedure involves adding the proper delta function to the intrinsic derivative. However, a different kind of delta function must be used: It must be one having infinite content. We call this kind of delta function a *super delta function* and its pulses *super delta pulses.*

In the preceding example only the intrinsic derivative of x^{-1} was needed; however, if we wish to know its complete derivative, it can be found by the use of standard differentiation procedures as shown in the following example.

Example B.

$$\frac{d}{dx}x^{-1} = \frac{d}{dx}|x|^{-1}\operatorname{sgn} x$$

$$= \operatorname{sgn} x\frac{d}{dx}|x|^{-1} + |x|^{-1}\frac{d}{dx}\operatorname{sgn} x$$

$$= -x^{-2} + 2\delta^2(x)$$

where $\delta^2(x) = |x|^{-1}\delta(x)$ is a super delta function.

The above procedure is generalized in the following theorem.

Theorem 9.1

Given: $F(x)$ is continuous except for an infinite discontinuity at $x = a$, i.e., $\beta = -1$. Multiple values of a are allowed.

$$\rightarrow\rightarrow\rightarrow \quad \frac{d}{dx}F(x) = \phi(x) + 2\operatorname{sgn} uF(x)\delta(u)\frac{du}{dx}$$

Where: **1.** $\phi(x)$ is the intrinsic derivative of $F(x)$.

2. $u = u(x)$ is any continuous function with $u(a) = 0$, $u(x) \neq 0$ when $x \neq a$, and with unit zero-crossing slope.

3. $\beta = \lim_{x\to a}\frac{F(x)}{F(2a - x)}$.

Proof of the theorem is given in Appendix A.

Some examples of the use of the theorem are given next. In some of these examples we have written the super delta function in the shorthand notation $2\delta^n(u)$, where

$$\delta^n(u) \equiv |u|^{-n+1}\delta(u) \tag{9.2}$$

This is an *nth-order delta function* and $\delta^n(u)$ is not necessarily the same as $[\delta(u)]^n$. That latter expression has not yet been defined.[1]

Example 9.1. $F(x) = x^{-m}$ where m is any positive odd integer, $u = x$, giving

$$\frac{d}{dx}x^{-m} = -mx^{-m-1} + 2\delta^{m+1}(x)$$

Example 9.2. $F(x) = \tan x$, $u = \cos x$, giving

$$\frac{d}{dx}\tan x = \sec^2 x - 2\delta^2(\cos x)$$

Example 9.3. $F(x) = \operatorname{ctn} x$, $u = \sin x$, giving

$$\frac{d}{dx}\operatorname{ctn} x = -\csc^2 x + 2\delta^2(\sin x)$$

[1]It is defined for $n > 0$ in Appendix B.

Example 9.4. $F(x) = \sec x$, $u = \cos x$, giving

$$\frac{d}{dx}\sec x = \sec x \tan x - 2\delta^2(\cos x)$$

Example 9.5. $F(x) = \csc x$, $u = \sin x$, giving

$$\frac{d}{dx}\csc x = -\csc x \operatorname{ctn} x - 2\delta^2(\sin x)$$

Example 9.6. $F(x) = \operatorname{sgn} x \operatorname{Ln}|x|$, $u = x$, giving

$$\frac{d}{dx}\operatorname{sgn} x \operatorname{Ln}|x| = |x|^{-1} - 2\,|\operatorname{Ln}|\,x\,\|\,\delta(x)$$

The functions $y = \operatorname{Ln}|x|$ and $y = \operatorname{sgn} x \operatorname{Ln}|x|$ are graphed in Figs. 9.3 and 9.4, respectively.

Note that not all super delta functions can be written in the form $\delta^n(x)$. An example of one that cannot is $|\operatorname{Ln}|\,x\,\|\,\delta(x)$ from Example 9.6. The integral of $|x|^{-1}$ is found from Example 9.6 to be

$$\int |x|^{-1}dx = \operatorname{sgn} x \operatorname{Ln}|x| + 2\int |\operatorname{Ln}|\,x\,\|\,\delta(x)\,dx \tag{9.3}$$

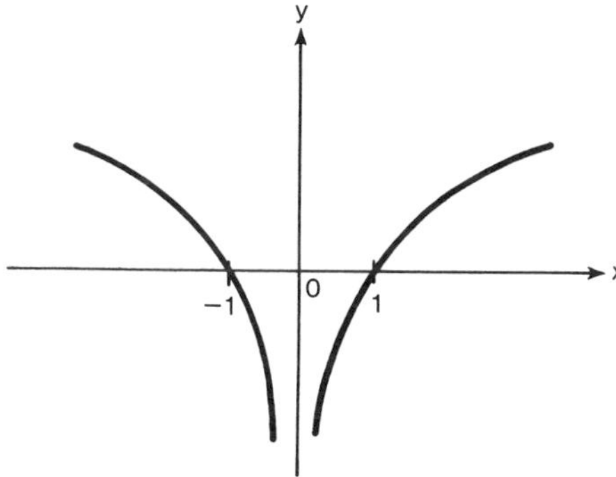

Figure 9.3 Graph of $y = \operatorname{Ln}|x|$.

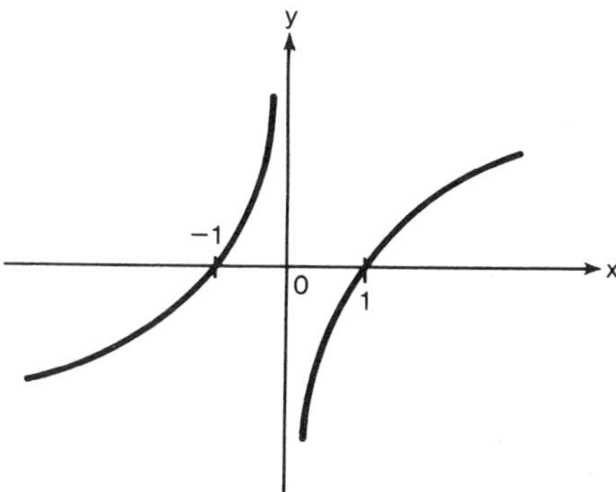

Figure 9.4 Graph of $y = \operatorname{sgn} x \operatorname{Ln}|x|$.

9.3 Integration

Continuous periodic functions such as the sine function can be integrated over any interval by use of the published integrals. However, periodic functions with singularities cannot generally be integrated across a singularity by use of the published integrals. An example of such a function is the secant-squared function. Its published integral, the tangent function, does not apply for intervals of integration that contain singularities.

The following paragraphs show how this situation can be remedied by the use of a super delta function.

Consider the definite integral

$$\int_a^b \sec^2 x \, dx = \tan x \Big|_a^b \tag{9.4}$$

It is apparent that if we let $a = 0$ and $b = \pi$, for example, that this equation gives the wrong answer, namely zero. The problem is that the integral function of $\sec^2 x$ is not $\tan x + C$; it is $\tan x + 2 \int \delta^2(\cos x) \, dx$, from Example 9.2. The function $y = \sec^2 x$ is graphed in Fig. 9.5.

The full significance is now clear of the claim made in a previous section that the published derivatives of tangent, cotangent, secant, and cosecant functions are not complete derivatives.

The complete expression for the definite integral of $\sec^2 x$ is

$$\rightarrow\rightarrow\rightarrow \quad \int_a^b \sec^2 x \, dx = \tan x \Big|_a^b + 2 \int_a^b \delta^2(\cos x) \, dx \tag{9.5}$$

Now if we let $a = 0$ and $b = \pi$, we obtain[2]

$$\int_0^\pi \sec^2 x \, dx = 2 \int_0^\pi \delta^2(\cos x) \, dx \tag{9.6}$$

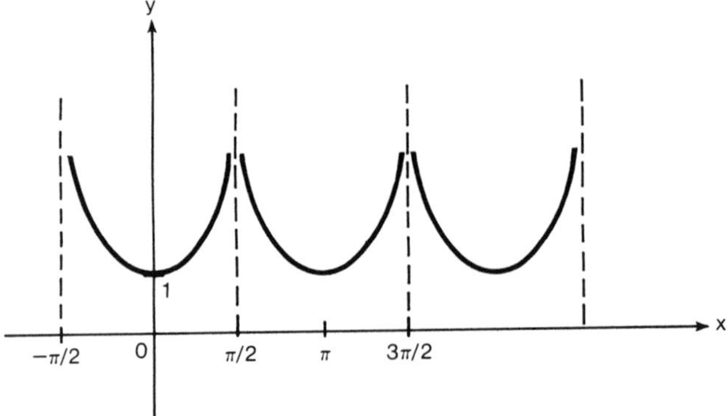

Figure 9.5 Graph of $y = \sec^2 x$.

[2] $\int_0^\pi \delta^2(\cos x)dx$ "evaluates" to $\delta(0)$. See Appendix B.

which is the content of a single super delta pulse: *viz., infinity*. This is a *particular infinity;* if the interval of integration is 0 to 2π, then the integral is *twice* that infinity because *two* super delta pulses are included in the interval of integration. These are the kinds of results that one might reasonably expect from those particular intervals of integration.

We stress that the super delta pulse train $\delta^2(\cos x)$ is effectively a bookkeeping tool.

The published integral of $1/x^3$ is

$$\int \frac{dx}{x^3} = -\tfrac{1}{2}x^{-2} + C \tag{9.7}$$

This is good for all x-values, including integration across $x = 0$ because $-2x^{-3}$ is the complete derivative of x^{-2}.

The published integral of $1/x$ is

$$\int \frac{dx}{x} = \text{Ln}\, x + C \qquad ((9.8))$$

This is good for all x-values, but it is not good for integration across zero because Ln x is real for positive x-values but complex for negative x-values. For all x-values including integration across $x = 0$, we use

$$\int \frac{dx}{x} = \text{Ln}\, |x| + C \tag{9.9}$$

To prove this, we write the complete derivative of Ln $|x|$, noting that Theorem 4.1 applies:

$$\frac{d}{dx}\text{Ln}\, |x| = \frac{1}{|x|}\,\text{sgn}\, x = 1/x \tag{9.10}$$

Since $1/x$ is the complete derivative of Ln$|x|$, it follows that Eq. (9.9) is good even across $x = 0$.

The published integral of $1/x^2$ is

$$\int \frac{dx}{x^2} = -x^{-1} + C \qquad ((9.11))$$

This is good for all x-values, but it is not good for integration across $x = 0$ because $-x^{-2}$ is not the complete derivative of x^{-1}. The complete expression for the integral function of $1/x^2$, from Example 9.1, is

$$\int \frac{dx}{x^2} = -x^{-1} + 2\int \delta^2(x)\, dx \tag{9.12}$$

This is good for all x-values including integration across $x = 0$. In general, where n is an even positive integer,

$$\rightarrow\rightarrow\rightarrow \quad \int \frac{dx}{x^n} = -\frac{1}{n-1}x^{-n+1} + \frac{2}{n-1}\int \delta^n(x)\,dx \qquad (9.13)$$

which is good for all x-values including integration across $x = 0$.[3]

In summary, for expressions of the form x^{-n}: **(1)** When n is an odd integer greater than 1, the published integrals are good for all x-values including integration across $x = 0$, **(2)** When n is an even (positive) integer, the published integrals are good for all x-values but they are not good for integration across $x = 0$, **(3)** For $n = 1$, Eq. (9.9) is good for all x-values including integration across $x = 0$, and **(4)** When n is an even (positive) integer, Eq. (9.13) is good for all x-values including integration across $x = 0$.

9.4 Summary of Some Uses for the Dirac Delta Function

Most of the functions in this chapter are functions of standard analysis. This is a vivid illustration of the fact that there is no sharp line of demarcation between relaxation analysis and standard analysis.

This chapter shows how the functions of standard analysis can be made to have increased usefulness by applying methods from relaxation analysis. Specifically, it shows how functions that could not previously be integrated across singularities now can be, with the help of delta functions.

Standard analysis has developed and survived over hundreds of years without the aid of delta functions, but relaxation analysis would not be possible without them. Now it seems that standard analysis, too, might benefit from their use.

Another important use for the Dirac delta function, in the form δ^0, is discussed in the following chapter. This application is to the production of *sections* through spaceforms. For example, cf. Eqs. (10.10).

[3]Such integrals are considered more fully in Chapter 11.

10

SPACEFORMS AND COMPUTER-GENERATED DISPLAYS

10.1 Classifying Spaceforms

This chapter presents mathematical relationships that are basic to the production of oscillographic, television, and computer-generated displays. We give a mathematical definition of spaceform. The discussion focuses on two- and three-dimensional spaceforms but is not limited to them. The mathematical relationships are presented and formulated in such a way that they can readily be extended to apply to four- and higher-dimensional spaceforms as well.

It is shown how the sawtooth wave and the zeroth-order delta function can be used to formulate mathematical descriptions of rasters, blanking and unblanking, and the production of sections and orthogonal projections of spaceforms.

The word *space* popularly conjures up the idea of a three-dimensional region. This narrow concept of space is replaced herein by the mathematical idea of an N-dimensional space, where N can be any positive integer greater than 1; indeed, even $N = 1$ leads to no apparent contradictions. The term *hyperspace* is also used to refer to spaces whose dimensionality is greater than three.

Just as the entirety of (Euclidean) two-dimensional space is called a *plane,* the entirety of three-dimensional space is called a *hyperplane* when speaking in the context of four-dimensional space. The term "hyperplane" is readily extended to apply to higher-dimensional spaces as a region of dimensionality one less than the dimensionality of the space.

The word *spaceform* as we use it is a general term for curves, surfaces, and "hypersurfaces." Two-dimensional spaceforms are of one kind: curves. Three-dimensional spaceforms are of two kinds: curves and surfaces. Four-dimensional spaceforms are of three kinds: curves, surfaces, and *3-flats*. N-dimensional ($N > 4$) spaceforms are of $N - 1$ kinds: curves, surfaces, 3-flats, . . ., ($N - 1$)-flats. M-flats ($M > 2$) are collectively called *hypersurfaces*. Spaceforms that can exist only in spaces with $N > 3$ are also called *hyperspaceforms*.

The above categorization of spaceforms is the mathematically traditional one. To those categories we add the *solids*. A 2-solid is the interior of a closed-plane curve. Points of the curve are *boundary points* of the 2-solid. An example is the interior of a circle—a *circular disc*. An example of a 3-solid is the interior of a sphere—a *spherical ball*. In general, an N-solid is the interior of a closed ($N - 1$)-flat.

We consider only Euclidean space; we need not concern ourselves with non-Euclidean spaces because these can be considered to be simply hypersurfaces that are embedded in a Euclidean space of one higher dimensionality.

10.2 Representing Spaceforms Mathematically

A group of two or more parametrically-related functions constitutes a *mathematical spaceform*. We write in general

$$\left.\begin{aligned} y_1 &= f_1(x) \\ y_2 &= f_2(x) \\ y_3 &= f_3(x) \\ &\cdots \\ y_N &= f_N(x) \end{aligned}\right\} \qquad (10.1)$$

The independent variable x is the *parameter;* the y_i ($i = 1, 2, 3, \ldots, N$) are the N-dependent variables. The number, N, of equations in the group is the dimensionality of the space in which the spaceform is embedded. It is also the dimensionality of the spaceform if the spaceform is *coordinate rectified,* i.e., positioned (rotated and translated) to minimize N. (An example of an unrectified spaceform is a circle that has been rotated into the third dimension, thereby requiring three parametric equations to describe this two-dimensional figure.) All spaceforms in this book are coordinate-rectified spaceforms unless otherwise indicated.

In the two-dimensional case we will usually take y_1 as the horizontal axis and y_2 as the vertical axis. In such a case we might use x in place of y_1 and y in place of y_2 provided there is no conflict with the parameter. Examples appear in Chapter 11.

In Eqs. (10.1), none of the f_i are constant because then the dimensionality of the spaceform would be less than N. Consequently, Eqs. (10.1) imply that y_j = const. where $j = N + 1, \ldots, \infty$. If we arbitrarily set one of the y_i = const., then we obtain the *orthogonal projection* of the N-dimensional spaceform onto the y_i = const. hyperplane.[1] This is an (N-1)-dimensional spaceform. We note that Eqs. (10.1) can also be written in matrix form as the vector

$$\begin{bmatrix} y_1 \\ y_2 \\ \cdots \\ y_N \end{bmatrix} = \begin{bmatrix} f_1(x) \\ f_2(x) \\ \cdots \\ f_N(x) \end{bmatrix} \tag{10.2}$$

If the functions of Eqs. (10.1) are all periodic and of commensurate periods (or frequencies), then the spaceform is repetitive.[2] (For two periods to be commensurate, their ratio must be a rational number. For all functions of a given spaceform to be of commensurate periods, their ratios of periods in all combinations must be rational numbers.)[3]

If one or more functions in the group are of incommensurate periods with any of the other functions, then the spaceform is nonrepetitive. If one or more functions are nonperiodic, then the spaceform is nonrepetitive.

10.3 Visualizing Spaceforms

Calling a group of parametric functions a spaceform implies the eventual ability to produce a visual object represented by the group of equations. This process of graphing or *spaceform visualization* concerns us in the present section.

Graphs of plane curves are routinely produced by hand, by computer, or by oscilloscope. Graphs of three-dimensional spaceforms can likewise be produced by hand (in models), by a holographic or stereo computer-

[1] A plane if $N = 3$, a line if $N = 2$. A hyperplane is named for the axis that it does not contain, for convenience.

[2] I.e., is traced repeatedly as x varies monotonically.

[3] When the ratio of the periods or frequencies of two periodic functions is a rational number, we say the waves are *synchronized;* if it is an irrational number, we say they are *unsynchronized.* If an oscillator is not synchronized with any other, we say it is *asynchronous* or *free running*. When referring to neural pulse trains, the term *entrained* is used instead of *synchronized.*

generated display, or by a three-dimensional counterpart of the oscilloscope such as the *parallactiscope*.[4]

Graphs of four-dimensonal spaceforms can be represented in sections or projections as a series of three-dimensional spaceforms produced by hand, by a computer, or by a parallactiscope. If we provide the capability to rotate the four-dimensional spaceform in hyperspace, then the observer can view the spaceform, so represented, from a variety of directions—as if he could "walk around it" in hyperspace. This we can do with dynamic real-time computer- and parallactiscope-generated displays.

Graphing of five- and higher-dimensional spaceforms presents additional challenges—not in kind, but in magnitude.

The patterns themselves, as presented on a real display device, must of necessity be limited to two and three dimensions. If the display medium is two dimensional, then two-dimensional spaceforms can be displayed in their entirety, whereas three-dimensional spaceforms must be displayed in sections or projections. If the display medium is three dimensional, then two- and three-dimensional spaceforms can be displayed in their entirety, whereas four-dimensional spaceforms must be displayed in sections or projections. Five-dimensional spaceforms can be doubly sectioned or projected for representation on a three-dimensional display medium, and so on.

Whether or not we can learn to "see" spaceforms of dimensionality greater than three is still open to debate. Since birth, we have been exposed to the three-dimensional objects that surround us. Thus we have learned to see them, that is, to perceive their forms. If we were constantly exposed to four-dimensional visual objects, could we learn to perceive their forms? Helmholtz is supposed to have believed that such a "hyperperceptive" ability could be developed if the brain is provided the proper inputs.[5] Manning[6] reported that he could "almost" see objects of hyperspace as a result of the study of four-dimensional geometry. Noll[7] reported on experiments involving the visualization of N-dimensional spaceforms by use of a digital computer. This writer[8] has likened the human

[4] H. B. Tilton, "Holoform Oscillography with a Parallactiscope," *1982 SID International Symposium Digest of Technical Papers,* 1st ed. (Coral Gables, Fla.: Lewis Winner, 1982), pp. 276–277.

[5] So reported by M. Gardner, "Mathematical Games," *Scientific American,* 215, no. 5 (November, 1966), 138–143.

[6] H. P. Manning, *Geometry of 4 Dimensions* (Dover Publications, 1956), p. 15.

[7] A. M. Noll, "A Computer Technique for Displaying N-dimensional Hyperobjects," *Communications of the ACM,* 10, no. 8 (August 1967), 469–473.

[8] H. B. Tilton, "Displays in Hyperspaceform Perception Experiments," *Journal of the Optical Society of America,* 68, no. 10 (October 1978), 1420.

observer to a defective four-dimensional observer, and has indicated that there is apparently no fundamental reason that representations of four-dimensional spaceforms cannot be displayed so as to convey all the information necessary for hyperspaceform perception.

But the potential (or lack of it) to develop a hyperperceptive ability has no bearing on the propriety or usefulness of referring to a group of four or more parametric equations as a *spaceform*. For certainly such a group of equations contains all the information necessary to define a visual object—regardless of whether or not such a visual object can be perceived by the human observer.

10.4 Two-Dimensional Spaceforms

10.4.1 General

Spaceforms that can exist in a plane are plane curves. Examples of these are given next.

An example of a nonrepetitive plane curve is

$$\left.\begin{aligned} y_1 &= x \\ y_2 &= x^2 \end{aligned}\right\} \tag{10.3}$$

This is a parabola. It is traced but once as we allow x to go from $-\infty$ to $+\infty$.

An example of a repetitive plane curve is

$$\left.\begin{aligned} y_1 &= A \sin x \\ y_2 &= B \cos x \end{aligned}\right\} \tag{10.4}$$

The reader will recognize Eqs. (10.4) as the parametric equations of an ellipse ($B \neq A$) or a circle ($B = A$), where A and B are constants.

The functions comprising a given spaceform are unique only if the arc length is to be a particular function of x. For example, the following duo of functions also defines a circle or ellipse:

$$\left.\begin{aligned} y_1 &= A \cos x \\ y_2 &= -B \sin x \end{aligned}\right\} \tag{10.5}$$

If A or B or both are functions of x so that they vary between zero and the upper limits A_0 and B_0 respectively, then points interior to the ellipse or circle are also defined by Eqs. (10.4). If all the interior points are so defined, then the spaceform is a 2-solid, namely, an elliptical or circular disc of size $y_1 = 2A_0$ by $y_2 = 2B_0$.

Now we consider

$$\left.\begin{aligned} y_1 &= A \sin (x/\tau_1) \\ y_2 &= B \cos (x/\tau_2) \end{aligned}\right\} \tag{10.6}$$

where τ_1 and τ_2 are numbers that are proportional to the periods of y_1 and y_2, respectively. If τ_1/τ_2 is an irrational number, then Eqs. (10.6) describe a "rectangular disc" or *plate* of size $y_1 = 2A$ by $y_2 = 2B$. We assume A and B to be constant.

If τ_1 and τ_2 are integers (or reciprocals of integers) then τ_1/τ_2 is a rational number, and if $\tau_2 \neq \tau_1$, then Eqs. (10.6) describe a two-dimensional *Lissajous curve*[9] or figure.[10] Such curves consist of two or more loops in the nature of a figure-8.

If we replace τ_1 and τ_2 with the smallest possible integers τ_1' and τ_2' so that $\tau_1'/\tau_2' = \tau_1/\tau_2$, then the number of loops in the y_1 direction is τ_1' and the number of loops in the y_2 direction is τ_2'. The form τ_1'/τ_2' is said to be the *least rational form* of the rational number τ_1/τ_2.

We note that Eqs. (10.6) do not give the most general form of two-dimensional Lissajous curves, as will become apparent when we discuss three-dimensional Lissajous curves.

10.4.2 Waveforms

The plane curve representation of a periodic function, such as the graph or image displayed on an oscilloscope, is referred to as a *waveform*.

If $p_a(x)$ is a periodic function with fundamental period a, then the plane curve defined by the duo of periodic functions

$$\left.\begin{aligned} y_1 &= A\ \mathrm{saw}_a(x/n) \\ y_2 &= B\ p_a(x) \end{aligned}\right\} \qquad (10.7)$$

is the waveform of $p_a(x)$. The number n is the number of cycles. The waveform of a periodic function is said to be *isomorphic over n cycles* to the function itself. The period of y_1 is proportional to n. When x is proportional to time, y_1 is called a *time base*.

Waveforms are unique among plane curves in that y_2 must be a single-valued function of y_1. Thus the "keystone waveform" shown in Fig. 10.1a is not really a waveform. Instead, it is a two-dimensional spaceform defined by two functions such as those graphed in Fig. 10.1b and Fig. 10.1c.

[9] Named after Jules Antoine Lissajous, who first demonstrated them in 1857.

[10] Of course, if $\tau_2 = \tau_1$, the curve is a circle or ellipse. Therefore we can, if we wish, consider a circle or ellipse to be a special case of a Lissajous curve. It is noted that the circle and ellipse are "conic sections" but the Lissajous curve (of more than one loop) is not. This may explain why Lissajous curves are not generally encountered in standard analysis.

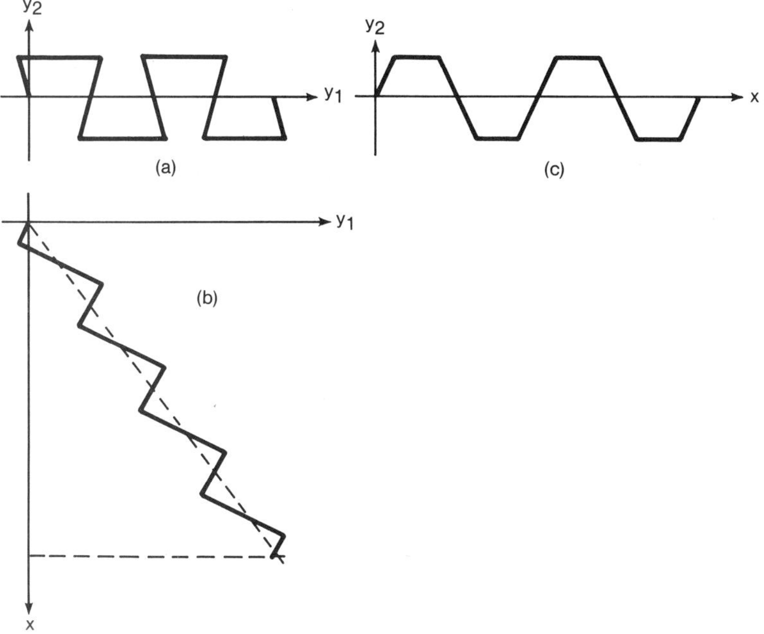

Figure 10.1 (a) A "keystone waveform"; (b) and (c) periodic functions which can generate the "keystone waveform."

10.4.3 Rasters

We now consider the plane curve

$$\left.\begin{aligned} y_1 &= A\ \mathrm{saw}_a(x\nu_1) \\ y_2 &= B\ \mathrm{saw}_a(x\nu_2) \end{aligned}\right\} \tag{10.8}$$

where ν_1 and ν_2 are numbers that are proportional to the frequencies of y_1 and y_2 respectively. If $1/\nu_1$ is an integer and $\nu_2 = 1$, this is just the waveform of $\mathrm{saw}_a x$ in $1/\nu_1$ cycles. If ν_1/ν_2 is an irrational number, then Eqs. (10.8) describe a plate of size $y_1 = aA$ by $y_2 = aB$. If $\nu_2 = \nu_1$, then a single diagonal of the plate is described.

If ν_1/ν_2 is a rational number, with ν_1'/ν_2' being the corresponding least rational form, then Eqs. (10.8) describe a two-dimensional *raster*. A raster consists of a series of parallel straight-line segments. If $\nu_1 >> \nu_2$, then the lines are nearly in the y_1 direction; the number of lines is ν_1', and ν_2' is the *interlace factor, m*.

If $\nu_2 >> \nu_1$, the lines are nearly in the y_2 direction; the number of lines is ν_2', and ν_1' is the interlace factor, m.

A single raster cycle constitutes a *frame*. There are said to be m

fields per frame, where m is the interlace factor. A field can be thought of as a "sub-frame" in that the number of lines in a field is 1/m times the number of lines in a frame, and the period of a field is 1/m times the period of a frame.

If $\nu_1 > \nu_2$, the period of a field is equal to the period of y_2. If $\nu_2 > \nu_1$, then the period of a field is equal to the period of y_1.

An important use for rasters is as the "canvas" upon which a television picture is "painted." American commercial television standards are based on $\nu_1/\nu_2 = 15\,750/60 = 525/2$, with y_1 being in the horizontal direction and y_2 being in the vertical direction, per Eqs. (10.8). Therefore the interlace factor in American television is 2, and there are two fields per frame. If a real television raster could use ideal sawtooth waves, there would be 525 lines in an American commercial television frame. Actually, the maximum number of lines that can be expected is about 340.[11]

10.5 How to Represent Shading and Intensity Mathematically

A concept that is useful in the applications is that of the *intensity of a point*.[12] A normal mathematical point is said to have an intensity of 1. The absence of a normal mathematical point is equivalent to the presence of a point whose intensity is zero. A continuum of intensities lying between those two limits is hypothesized. Thus, the picture that is painted on a two-dimensional raster in television can be represented in terms of points of varying intensity along the lines that compose the raster.

The information in a complete monochromatic television picture can be represented by the following three functions constituting a two-dimensional *shaded spaceform:*

$$\left.\begin{aligned} y_1 &= A \operatorname{saw}_a(x\nu_1) \\ y_2 &= B \operatorname{saw}_a(x\nu_2) \\ z &= 1 - z'(x) \end{aligned}\right\} \qquad (10.9)$$

Here, y_1 and y_2 define the raster and z defines the point-to-point intensity. If a still picture is displayed, then z is periodic with period equal to the frame period. Trichromatic television requires three z-functions.

When $z = 0$ (or less), the point is said to be *blanked*. When $z = 1$

[11] D. G. Fink, ed., *Television Engineering Handbook*, 1st ed. (New York: McGraw-Hill, 1957), pp. 1–13.

[12] This use of the word "intensity" is different from the photometric one, although they are conceptually similar.

(or more), the point is said to be *unblanked*. When $0 < z < 1$, the point is said to be *shaded*. In Eqs. (10.9), z is the *intensity function* and z' is the *shading function*. The functions z and z' are 1's complements. If z' and z are allowed to assume only the two values 0 and 1, then they are further categorized as *blanking* and *unblanking* functions, respectively.[13]

Equations (10.9) can also be thought of as a three-dimensional spaceform. In this way, the "intensity of a point" can be treated in more conventional terms, as simply another space coordinate—a "z-axis." Hence the term "z-axis modulation," which is sometimes encountered in oscillography when referring to intensity modulation.

The converse relationship is also useful, i.e., a space coordinate can sometimes be treated as an intensity relationship. This results in a shaded spaceform of dimensionality one less than the original.

Another use for the intensity function is in producing *sections* through spaceforms. For example, the *section through the N-dimensional spaceform* of Eqs. (10.1) *in the hyperplane*[14] $y_i = A = const.$ is[15]

$$\left.\begin{aligned} y_1 &= f_1(x) \\ y_2 &= f_2(x) \\ &\cdots \\ y_N &= f_N(x) \\ z &= \delta^0[f_i(x) - A] \end{aligned}\right\} \quad (10.10)$$

Thus when $f_i(x) = A$, $z = 1$, and when $f_i(x) \neq A$, $z = 0$.

The dimensionality of the spaceform of Eqs. (10.10) is $N - 1$ for the following reason. The y_i coordinate information has effectively collapsed to $y_i = A$ because when it is not A, the spaceform is blanked. Therefore we can, if we wish, replace the $y_i = f_i(x)$ component of the spaceform with $y_i = A$ without changing the spaceform.

Examples of the use of Eqs. (10.10) are given in Chapter 11.

10.6 Rasterforms

Any plane curve can be approximated as a two-dimensional *rasterform*. This is a representation of a spaceform that is painted on a raster by letting z assume only the two values 0 and 1. It is an approximation

[13] In the context of xy recorders, z' is a *pen lift* function.

[14] A plane if $N = 3$, a line if $N = 2$.

[15] Refer to Eqs. (2.49) for the meaning of δ^0.

because a raster is composed of a discrete number of lines, i.e., it is not "space filling."

In the applications, z is most generally an irregular train of pulses that repeats with period equal to the frame period of the raster. This produces a repetitive representation of an unchanging spaceform.

The z-function is determined by noting the coincidence between the raster and the curve to be represented as a rasterform. Where coincidence exists, $z = 1$; where it does not, $z = 0$. The following paragraph details these relationships.

Let the raster be defined by Eqs. (10.8) and let the curve to be represented as a rasterform be defined by

$$\left.\begin{aligned} y_1' &= f_1(x') \\ y_2' &= f_2(x') \end{aligned}\right\} \qquad (10.11)$$

where the parameter x' is independent of x in Eqs. (10.8). The desired unblanking function can be written

$$z = \delta^0(y_1 - y_1')\delta^0(y_2 - y_2') \qquad (10.12)$$

where y_1 and y_2 refer to Eqs. (10.8), and y_1' and y_2' refer to Eqs. (10.11). The situation is diagrammed in Fig. 10.2. In this application δ^0 acts as a coincidence detector as discussed in Chapter 2.

By using Eq. (10.12), rasterforms can be generated in real time for high-frequency[16] repetitive spaceforms. This approach is not practical for use with low-frequency spaceforms. For this latter case, the two functions representing the spaceform can first be stored in memory, then read out to form the unblanking function. Thus instead of Eq. (10.12), z would be given by a relationship of the form

$$z = Y'(y_1, y_2) \qquad (10.13)$$

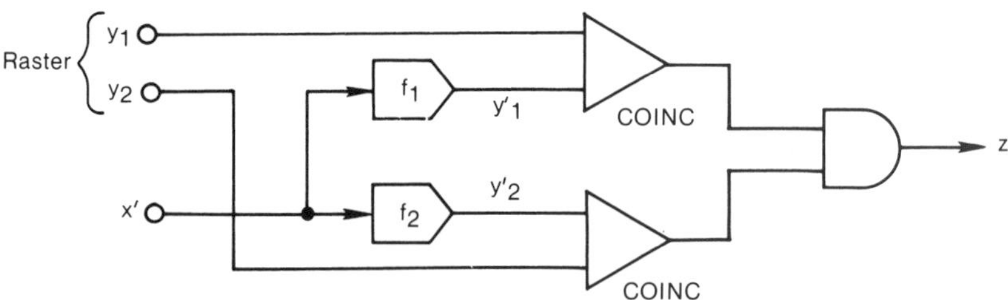

Figure 10.2 Block diagram of Eq. (10.12).

[16] High relative to the raster generating ("scanning") frequencies.

Thus, each point of the raster would have either a ZERO or a ONE associated with it, depending on the particular spaceform Y'. As the raster is generated, the spaceform appears upon it, so that the entire spaceform is painted once each frame.

Two methods of "analog storage" based on this approach are discussed next. In the first, we assume that $y_2' = F(y_1')$ can be found from Eqs. (10.11), and that it is a single-valued function. In this form we note that

$$y_2' - F(y_1') = 0 \tag{10.14}$$

represents all points of the spaceform, and that

$$y_2' - F(y_1') \neq 0 \tag{10.15}$$

represents all points not of the spaceform. It follows that the desired unblanking function is

$$z = \delta^0[y_2' - F(y_1')] \tag{10.16}$$

where $y_1' = y_1$ and $y_2' = y_2$, with y_1 and y_2 being raster-point coordinates from Eqs. (10.8), so that finally

$$z = \delta^0[y_2 - F(y_1)] \tag{10.17}$$

The block diagram for this case is shown in Fig. 10.3.

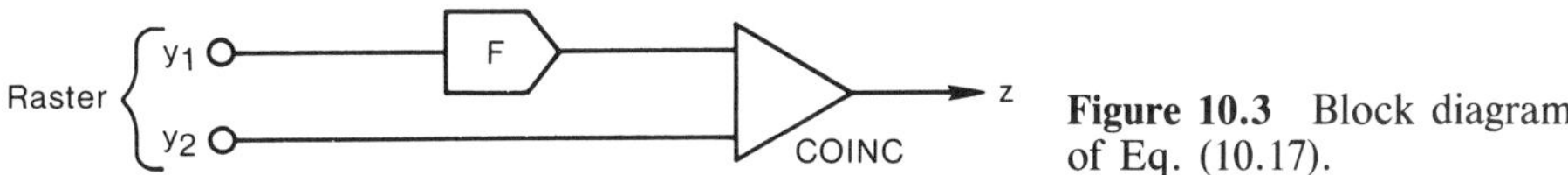

Figure 10.3 Block diagram of Eq. (10.17).

If y_2' is a multiple-valued function of y_1', another method is available as follows. We write the spaceform Eqs. (10.11) in the following form:

$$\left.\begin{aligned} x_1' &= g_1(y_1') \\ x_2' &= g_2(y_2') \end{aligned}\right\} \tag{10.18}$$

where g_1 is the inverse function to f_1 and g_2 is the inverse function to f_2. For the moment we assume that both functions in Eqs. (10.18) are single valued. We note that

$$x_2' - x_1' = 0 \tag{10.19}$$

represents all points of the spaceform, and that

$$x_2' - x_1' \neq 0 \tag{10.20}$$

represents all points not of the spaceform. It follows that the desired unblanking function is

$$z = \delta^0[g_2(y_2) - g_1(y_1)] \tag{10.21}$$

The block diagram for this case is shown in Fig. 10.4.

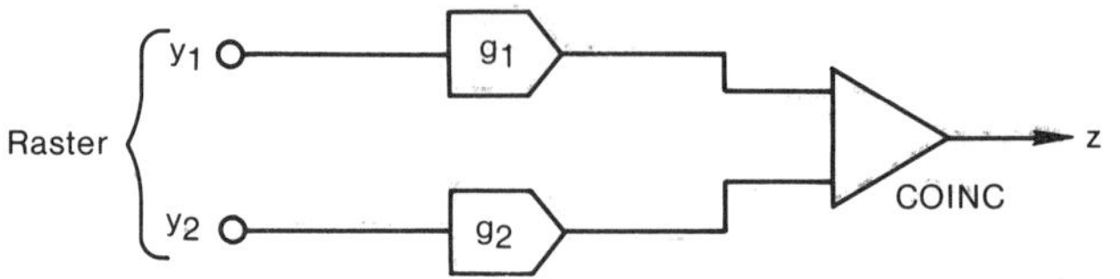

Figure 10.4 Block diagram of Eq. (10.21).

Two examples follow that show how the two methods can be extended and used with multiple-valued functions. The spaceform chosen is a circle. The first example is based on the first method and the second example is based on the second method.

Example 10.1. Generating the rasterform of a circle by use of analog storage of the spaceform using first method. Desired spaceform is

$$y_1 = \cos x$$

$$y_2 = \sin x$$

These equations can be solved for y_2 in terms of y_1 to give

$$y_2 = \pm(1 - y_1^2)^{1/2}$$

We require a function generator whose output, u, is

$$u = (1 - y_1^2)^{1/2}, \quad -1 < y_1 < 1$$
$$= 0 \text{ elsewhere.}$$

We generate two z-functions based on the block diagram of Fig. 10.5: one to paint the upper semicircle and the other to paint the lower one. The two z-functions can be OR'ed to give the final, desired unblanking function.

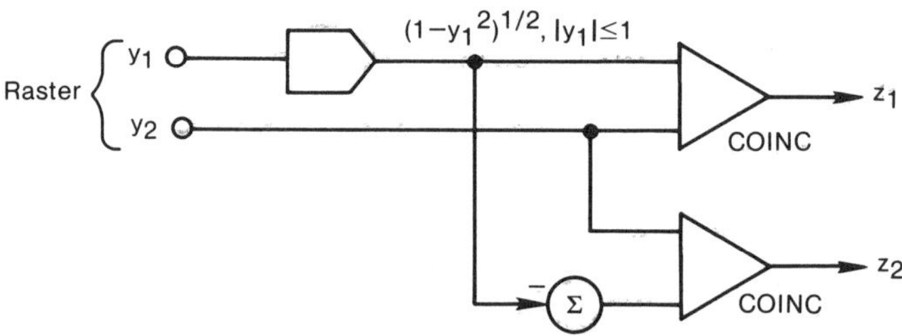

Figure 10.5 Block diagram for the rasterform circle generator of Example 10.1.

Example 10.2. Generating the rasterform of a circle by use of analog storage of the spaceform using second method.

We write the spaceform equations in the form

$$x_1 = \mathrm{Cos}^{-1} y_1$$

$$x_2 = \mathrm{Sin}^{-1} y_2$$

But Cos^{-1} and Sin^{-1} do not cover the same range of values. We begin by limiting the range of values for both to the range 0 to $\frac{1}{2}\pi$. This is done by limiting y_1 and y_2 to the range from 0 to $+1$. It is not difficult to show that this will cause the quarter-circle that lies in the first quadrant to be generated. Further consideration shows that in order to generate the remaining portions of the circle we simply need to write

$$x_1 = \text{Cos}^{-1}|y_1|$$

$$x_2 = \text{Sin}^{-1}|y_2|$$

The block diagram for this case is shown in Fig. 10.6. Again, the outputs of the function generators need to be zero when the inputs are outside the range from -1 to $+1$.

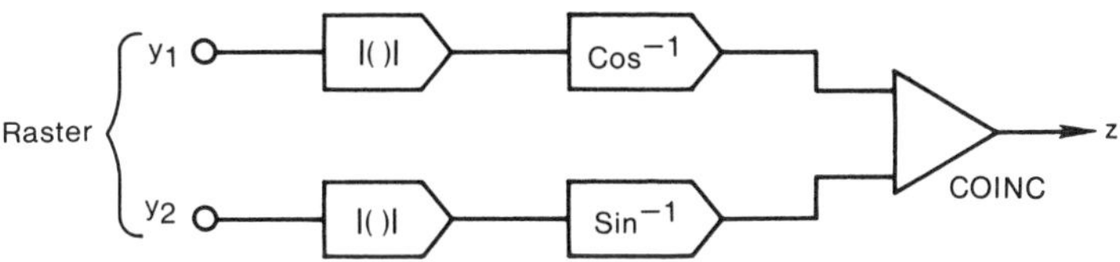

Figure 10.6 Block diagram for the rasterform circle generator of Example 10.2.

A method that enjoys more widespread use than the methods of analog storage is based on digital storage. This is a versatile method, especially if a programmable digital memory is used. Indeed, such a system can be designed to operate in near-real time. These methods are well-known to the digital art, and a discussion of them is clearly outside the scope of this volume.

10.7 Three-Dimensional Spaceforms

10.7.1 General

Three-dimensional spaceforms can be coded onto a plane as orthographic, oblique, or scenographic (perspective) projections. Thus they become, in fact, two-dimensional. But beyond this, truly three-dimensional spaceforms exist as twisted curves and surfaces in three-dimensional (and higher-dimensional) spaces only. Examples of some of these follow.

10.7.2 Helixes and cylinders

If we take the two functions of Eq. (10.4) and include a sawtooth wave of period commensurate with the sine and cosine, then the circle or ellipse is stretched into a helix—a coil-spring-shaped curve. Equations for this twisted curve are

$$\left.\begin{aligned} y_1 &= A \sin x \\ y_2 &= B \cos x \\ y_3 &= C \operatorname{saw}_{2\pi}(x/n) \end{aligned}\right\} \quad (10.22)$$

where n is the number of turns in the helix if n is an integer.

If $n = a/b$ is a rational fraction, and a'/b' is the corresponding least rational fraction, then Eqs. (10.22) describe a multiple helix where a' is the number of turns (total number per helix cycle) and b' is the interlace factor. The double helix of DNA has an interlace factor of 2, for example.

Finally, if n is an irrational number, then Eqs. (10.22) describe a right elliptical (or circular) cylinder of size $y_1 = 2A$, $y_2 = 2B$, and $y_3 = 2\pi C$.

Plate 1 shows a series of photos of a helix taken from a parallactiscope display. The camera was moved between photos to capture the image from different directions. By viewing adjacent photos with a stereoscopic viewer (noncrossed disparity), stereo depth can be seen. When viewing the live parallactiscope displays, stereo depth and movement parallax are obtained automatically without the use of a viewer.

Plate 2 shows a double helix, as described above (interlace factor of two).

10.7.3 Three-dimensional Lissajous curves

Three-dimensional Lissajous curves, by analogy with two-dimensional ones, are defined by three sinusoidal waves. We write in general, with all three functions of commensurate periods,

$$\left.\begin{aligned} y_1 &= A \sin (x/\tau_1 + \phi_1) \\ y_2 &= B \sin (x/\tau_2 + \phi_2) \\ y_3 &= C \sin (x/\tau_3 + \phi_3) \end{aligned}\right\} \quad (10.23)$$

where the ϕ_i ($i = 1, 2, 3$) are the phases. For example, Eqs. (10.6) can be obtained from Eqs. (10.23) by letting $\phi_1 = 0$, $\phi_2 = \frac{1}{2}\pi$, and $C = 0$.

The number of loops in the y_1 direction is τ_1'; in the y_2 direction, τ_2'; and in the y_3 direction, τ_3', where τ_1', τ_2', and τ_3' are the smallest integers that simultaneously satisfy the three conditions $\tau_1'/\tau_2' = \tau_1/\tau_2$, $\tau_1'/\tau_3' = \tau_1/\tau_3$, and $\tau_2'/\tau_3' = \tau_2/\tau_3$.

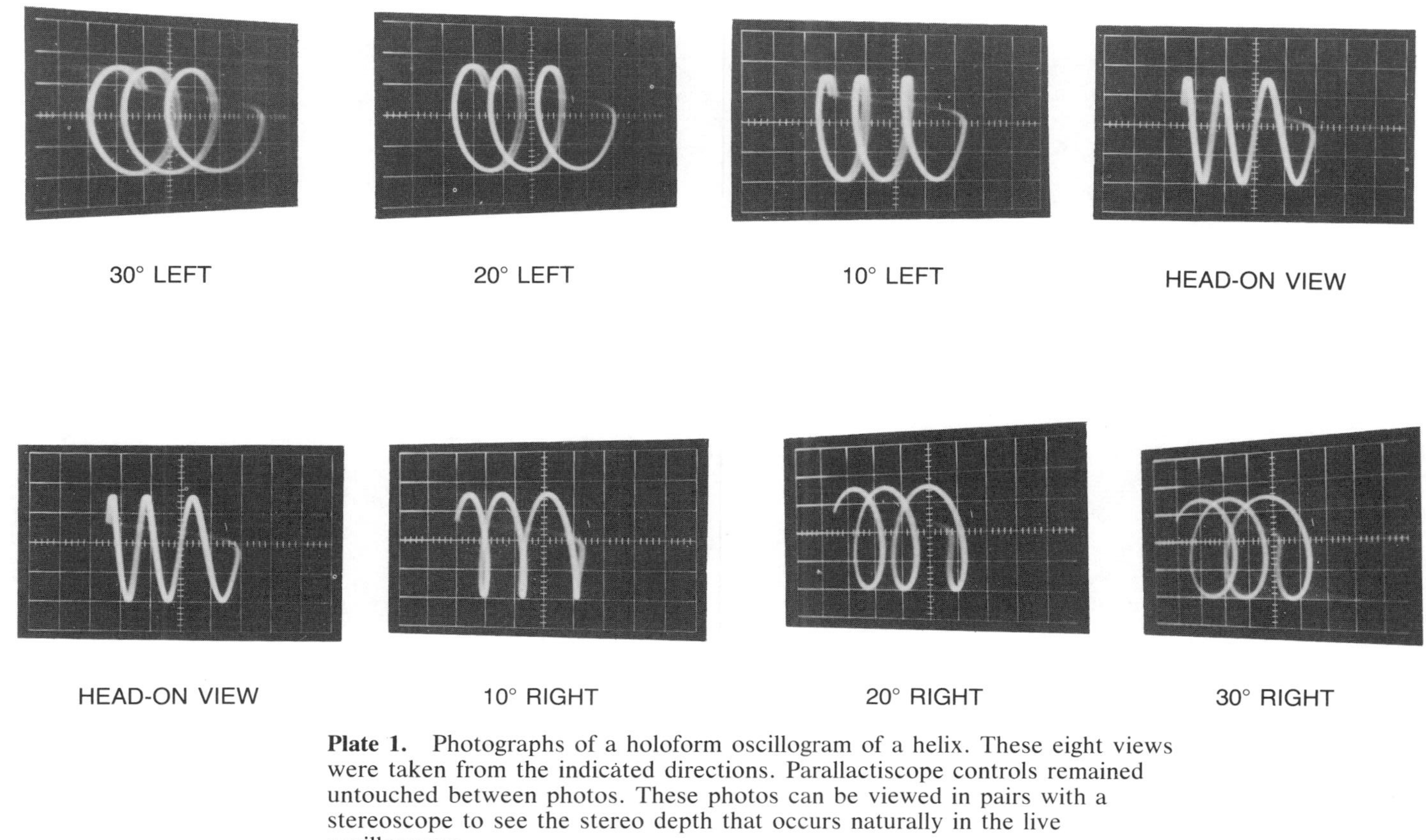

Plate 1. Photographs of a holoform oscillogram of a helix. These eight views were taken from the indicated directions. Parallactiscope controls remained untouched between photos. These photos can be viewed in pairs with a stereoscope to see the stereo depth that occurs naturally in the live oscillograms.

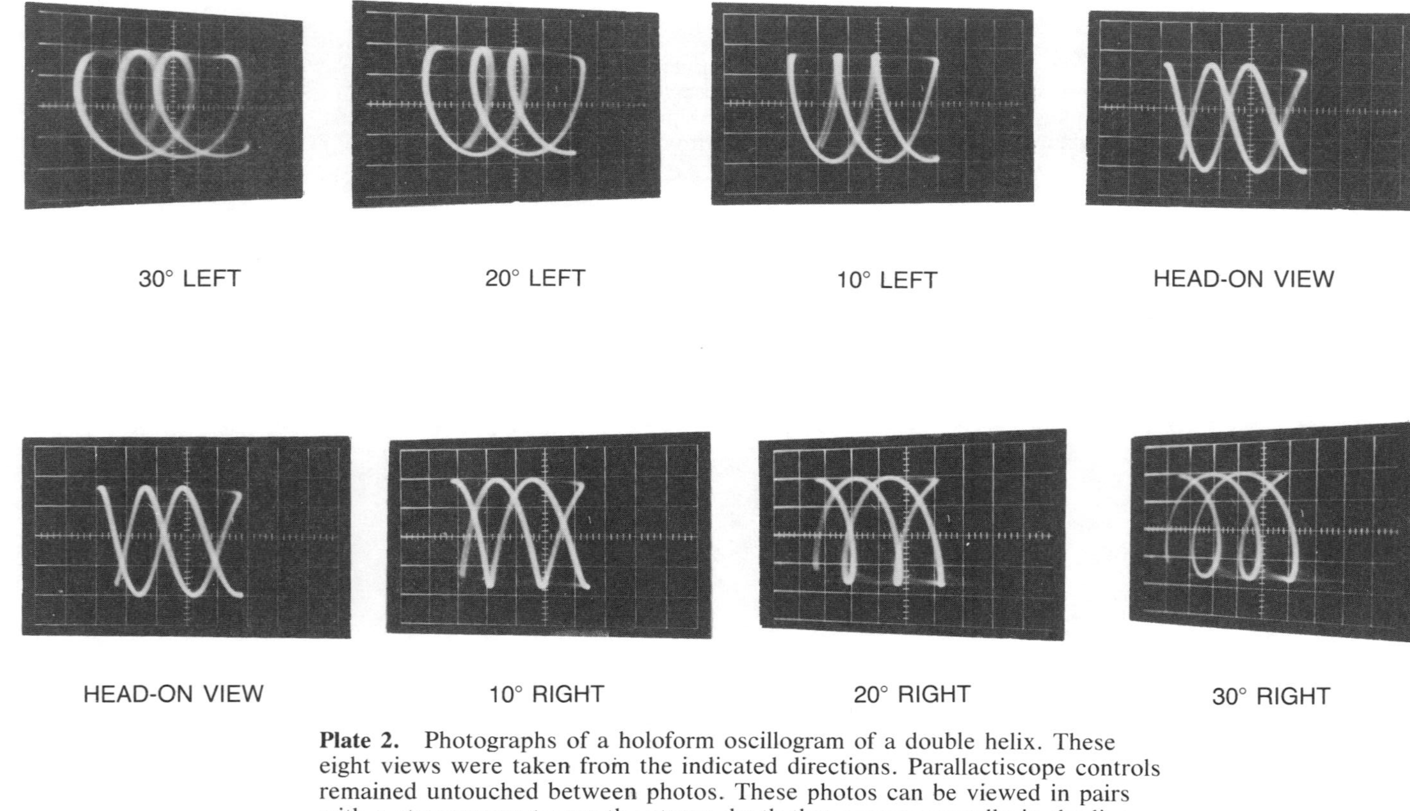

Plate 2. Photographs of a holoform oscillogram of a double helix. These eight views were taken from the indicated directions. Parallactiscope controls remained untouched between photos. These photos can be viewed in pairs with a stereoscope to see the stereo depth that occurs naturally in the live oscillograms.

10.7.4 Three-dimensional rasters

A logical extension of Eqs. (10.8) to the case of three dimensions is, with all three functions being of commensurate frequencies,

$$\left.\begin{aligned} y_1 &= A\ \mathrm{saw}_a(x\nu_1) \\ y_2 &= B\ \mathrm{saw}_a(x\nu_2) \\ y_3 &= C\ \mathrm{saw}_a(x\nu_3) \end{aligned}\right\} \quad (10.24)$$

Let ν_1', ν_2', ν_3' be the smallest integers that simultaneously satisfy the three conditions $\nu_1'/\nu_2' = \nu_1/\nu_2$, $\nu_1'/\nu_3' = \nu_1/\nu_3$, and $\nu_2'/\nu_3' = \nu_2/\nu_3$.

If $\nu_1 >> \nu_2 >> \nu_3$, then the lines comprising the raster are nearly in the y_1 direction, and the number of lines is ν_1'. There are three interlace factors that can be found by taking the functions in pairs and analyzing as for the two-dimensional case.

For example, let $\nu_1' = 15\,750$, $\nu_2' = 60$, and $\nu_3' = 1$. The number of lines is 15 750 (these are nearly in the y_1 direction), and the three interlace factors are 2, 1, 1.

Any three-dimensional spaceform can be approximated as a three-dimensional rasterform in a manner parallel to that for the two-dimensional case.

10.8 The Hidden-Line Problem

Three-dimensional spaceforms as displayed in computer- and parallactiscope-generated displays are inherently transparent because points "show through" intervening portions of the spaceform. The effect is especially noticeable when only edges and outlines of surfaces and solids are drawn. These are so-called "wire frame" displays.

If we desire to display opaque visual objects, a blanking function must be specified. This function must be constituted in such a way that it blanks those points that are "behind" other obscuring portions of the visual object. The problem of specifying such a function is referred to as the *hidden-line problem*.

Computer- and parallactiscope-generated displays are "friendly" to the solution of the hidden-line problem. That is, a solution is possible without placing restrictions on the observer's viewing position. Not all real-time three-dimensional display devices are friendly to the solution of the hidden-line problem.

The hidden-line problem cannot be readily solved for any but the simplest spaceforms by using analog methods. There are, however, ways to approximate the required blanking function wih a shading function.[17]

The hidden-line problem can be exactly solved using digital techniques.

10.9 An Electronic Application of Three-Dimensional Space

The characteristic surface of an analog multiplier can be visualized by the use of a three-dimensional display. Its mathematical description is

$$\left.\begin{aligned} y_1 &= u \\ y_2 &= u' \\ y_3 &= Auu' \end{aligned}\right\} \quad (10.25)$$

where u and u' are multiplier inputs (independent variables), A is a constant, and y_3 is multiplier output (dependent variable). For example, $u = \sin \omega t$ and $u' = \sin \omega' t$ where $\omega' >> \omega$. The resulting characteristic saddle-shaped surface (hyperbolic paraboloid) of this device is shown in the series of parallactiscope photos in Plate 3 as viewed from various angles.

By viewing live displays of this type one can get a rapid insight into a given multiplier's worth in terms of dynamic range, limits of frequency response, and its "linearity." Dynamic range can be explored by increasing the input signal levels and noting the point at which saturation occurs. Usable frequency limits can be explored by noting the frequency at which output begins to drop off. Phase-shift effects produce a hysteresis loop characteristic in the display. Linearity is explored by holding one input constant and varying the other. Under such conditions, an analog multiplier operates as a linear device.

Sheingold[18] also discusses the characteristic surface of an analog multiplier.

10.10 Four-Dimensional Spaceforms (Why stop at three?)

Four-dimensional Lissajous curves and rasters can easily be formulated as logical extensions of the three-dimensional cases. Thus four-dimensional

[17] See H. B. Tilton, "Real-Time Direct-Viewed CRT Displays Having Holographic Properties," *Proceedings of the Technical Program, Electro-Optical Systems Design Conference—1971 West,* (Chicago: Industrial and Scientific Conference Management, Inc., 1971), pp. 415–422. See the appendix on page 422.

[18] D. H. Sheingold, ed., *Nonlinear Circuits Handbook,* 2nd ed., by the Engineering Staff of Analog Devices, Inc. (Norwood, Mass.: Analog Devices, Inc., January 1976). See page 207.

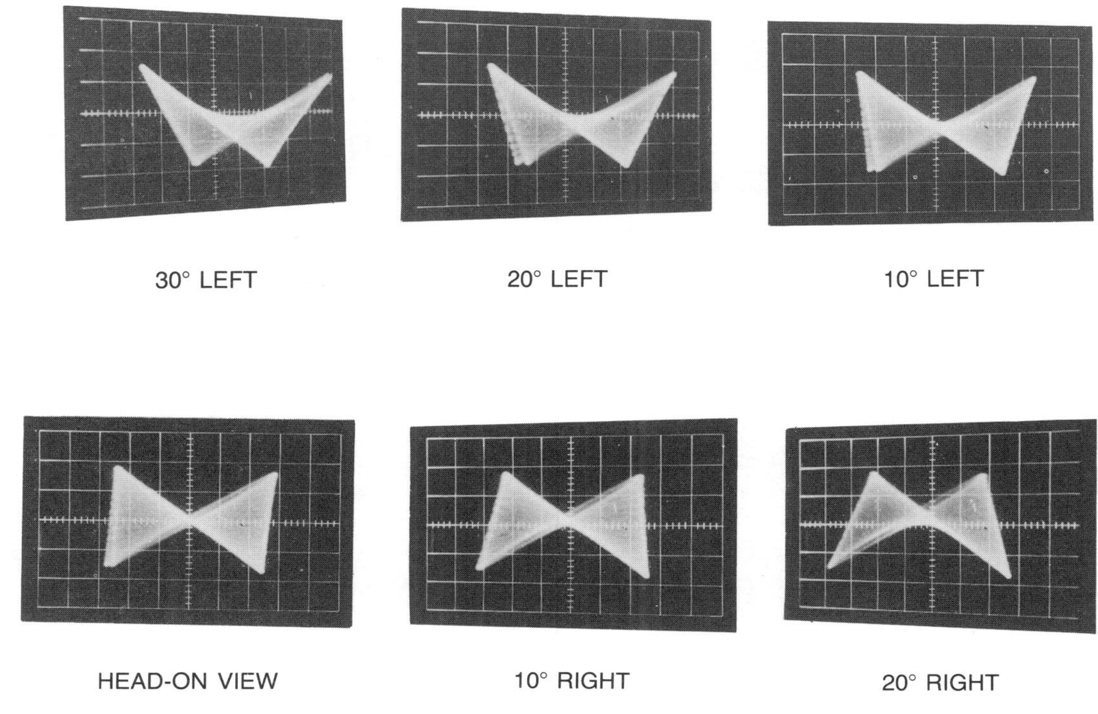

Plate 3. Photographs of a holoform oscillogram of the characteristic surface of an analog multiplier, a Burr-Brown 4213. Horizontal and depth deflections are multiplier inputs. Vertical deflection is multiplier output. This surface is a hyperbolic paraboloid saddle-shaped surface. It is a ruled surface. View with a stereoscope, as described in Plates 1 and 2.

Lissajous curves would use y_1, y_2, and y_3 from Eqs. (10.23) along with the following y_4:

$$y_4 = D \sin (x/\tau_4 + \phi_4) \tag{10.26}$$

and four-dimensional rasters would use y_1, y_2, and y_3 from Eqs. (10.24) along with the following y_4:

$$y_4 = D \operatorname{saw}_a(x\nu_4) \tag{10.27}$$

A classic figure of hyperspace, the *tesseract, hypercube,* or *four-dimensional orthotope,* is described in outline by the parametric equations

$$\left.\begin{aligned} y_1 &= A \operatorname{orth}_4 x \\ y_2 &= A \operatorname{orth}_4(x - \pi/2) \\ y_3 &= A \operatorname{orth}_4(x - \pi) \\ y_4 &= A \operatorname{orth}_4(x - 3\pi/2) \end{aligned}\right\} \tag{10.28}$$

where $\operatorname{orth}_4 x$ is as shown in Fig. 10.7. Equations (10.28) repetitively define the 32 edges of the tesseract. Any three of those functions define the edges of a cube, and any two define the edges of a square.

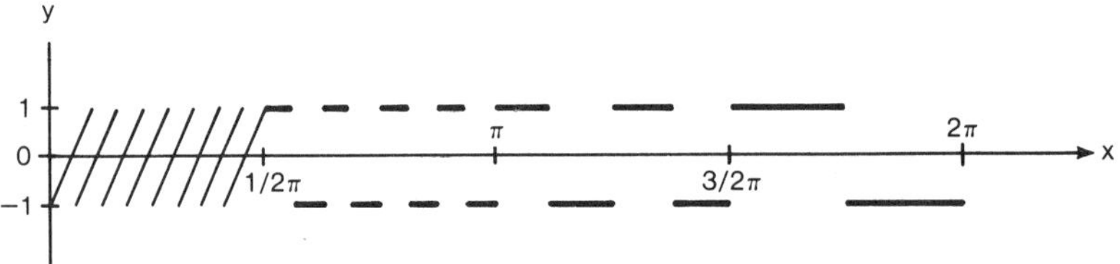

Figure 10.7 The function $\operatorname{orth}_4 x$ of Eq. (10.28).

An application of four-dimensional space was suggested independently by D'Alembert and Lagrange wherein time is the fourth dimension. An example of this is the *space-time continuum* of relativity. Another application concerns the representation of functions of a complex variable. This application was suggested by Poincaré. An example of this application is given in Chapter 11.

An application of five-dimensional space that suggests itself, in light of the above, is to time-varying functions of a complex variable wherein time is the fifth dimension.

In the next chapter these ideas are used to show how functions involving complex variables can be represented as three- and four-dimensional spaceforms; and how these can be visualized for some of the more common transcendental functions. It is shown how representation of these functions in terms of real and imaginary parts is facilitated by the use of relaxation functions.

11

COMPLEX FUNCTIONS AS SPACEFORMS

11.1 A Review of Complex Notation

The complex variable $\mathbf{x}$ can be written in rectangular form as

$$\mathbf{x} = x + ix' \tag{11.1}$$

where x and x' are real variables and $i^2 = -1$. Here x is the real part of $\mathbf{x}$, or $x = \operatorname{Re} \mathbf{x}$; and x' is the coefficient of the imaginary part of $\mathbf{x}$, or $x' = \operatorname{Im} \mathbf{x}$.[1] Consequently, Eq. (11.1) can also be written

$$\mathbf{x} = \operatorname{Re} \mathbf{x} + i \operatorname{Im} \mathbf{x} \tag{11.2}$$

Any complex number can be represented as a point in the complex number plane as shown in Fig. 11.1. Such a diagram is called an *Argand diagram*. The complex number plane is isomorphic with the plane defined by the rectangular Cartesian coordinate system x, x'. Consequently, the point at (x, x')—and hence any complex number—can be represented in terms of polar coordinates r, θ in matrix form as the vector

$$\mathbf{x} = \begin{bmatrix} x \\ x' \end{bmatrix} = \begin{bmatrix} r \cos \theta \\ r \sin \theta \end{bmatrix} \tag{11.3}$$

where r is the *Argand radius* or simply *radius* of $\mathbf{x}$, with $r = (x^2 + x'^2)^{1/2}$. The definition of absolute value is traditionally generalized to mean

[1] When $\operatorname{Im} \mathbf{x} = 0$, $\mathbf{x}$ becomes x; so that the real variable x is just a special case of the complex variable $\mathbf{x}$.

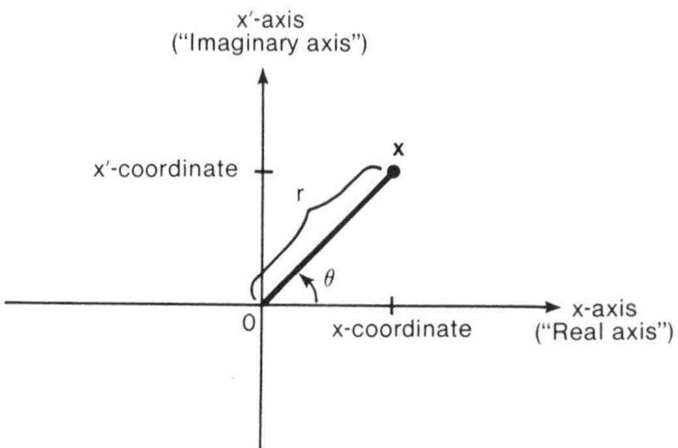

Figure 11.1 Argand diagram.

$$|x + ix'| = (x^2 + x'^2)^{1/2} \tag{11.4}$$

Therefore, r is also the absolute value or *magnitude* of **x**, or[2]

$$r = |\mathbf{x}| \tag{11.5}$$

The angle θ can assume any value between $-\infty$ and ∞, so that any given complex number has an infinite number of θ-values associated with it. That is, θ is multiple valued, and any complex number is multiple valued in terms of θ. It can be rendered single valued by substituting Θ for θ, where Θ is the principal value of θ defined in Appendix D. We also write arg **x** for θ and Arg **x** for Θ. Just as $|\mathbf{x}|$ is the Argand radius of **x**, Arg **x** is the *Argand angle* or simply *angle* of **x**.[3] Although not universally agreed on, Θ will be taken to range between $-\pi$ and π when it is to span a range of 2π.[4]

It is customary to write Θ as Arctan(x'/x), but this requires a special definition for Arctan if Θ is to span a range of 2π. Instead, it is preferable to write Θ so that Tan^{-1} can be used as previously defined. Three methods for doing this are developed in Appendix D. Using one method, Θ can be written

$$\Theta = 2\,\text{Tan}^{-1}[x'/(r + x)] \tag{11.6}$$

Using the second method, Θ can be written

$$\Theta = 2\,\text{Tan}^{-1}[(r - x)/x'] \tag{11.7}$$

Using the third method, Θ can be written

$$\Theta = \text{Tan}^{-1}(x'/x) + \tfrac{1}{2}\pi \,\text{sgn}\, x' - \tfrac{1}{2}\pi \,\text{sgn}(xx') \tag{11.8}$$

[2] $|\mathbf{x}|$ is also called the *modulus* or *norm* of **x**. Another symbol encountered in the literature is $\|\mathbf{x}\|$.

[3] Arg **x** is also called the *argument* or *amplitude* of **x**, not to be confused with previous usage of these two words. Other symbols encountered in the literature are Amp **x** and ∢**x**.

[4] Sometimes even a range of 2π is too wide. For example, if $\Theta = \text{Sin}^{-1}(\ )$, then Θ spans a range of only π.

It is clear from Eq. (11.3) that **x** can be written in trigonometric form as

$$\mathbf{x} = r(\cos\theta + i\sin\theta) \tag{11.9a}$$

$$\mathbf{x} = r(\cos\Theta + i\sin\Theta) \tag{11.9b}$$

The infinite series expansion for $\cos\theta + i\sin\theta$ is identical to that for $e^{i\theta}$. Therefore, another way to write the complex variable **x** is in exponential form as

$$\mathbf{x} = r\, e^{i\theta} \tag{11.10a}$$

$$\mathbf{x} = r\, e^{i\Theta} \tag{11.10b}$$

Other notations found in the literature for $e^{i\theta} = \cos\theta + i\sin\theta$ are $\exp(i\theta)$, cis θ (from $\underline{c}$os θ + $\underline{i}$ $\underline{s}$in $\underline{\theta}$), and $\angle\theta$. Later, we find that $\operatorname{sgn}\mathbf{x} = e^{i\theta}$ as well.

The sum of two complex quantities **x** and **y** is, following the usual rules of addition,

$$\mathbf{x} + \mathbf{y} = (\operatorname{Re}\mathbf{x} + \operatorname{Re}\mathbf{y}) + i\,(\operatorname{Im}\mathbf{x} + \operatorname{Im}\mathbf{y}) \tag{11.11}$$

Their product is

$$\mathbf{xy} = |\mathbf{x}||\mathbf{y}|e^{i(\arg\mathbf{x} + \arg\mathbf{y})} \tag{11.12}$$

We can also multiply in rectangular form, following the usual rules of multiplication with $i^2 = -1$, but we cannot add in exponential form (except to obtain $\mathbf{x} + \mathbf{y} = |\mathbf{x}|e^{i\arg\mathbf{x}} + |\mathbf{y}|e^{i\arg\mathbf{y}}$). We also note that $1/i = -i$.

In exponential form the reciprocal of **x** is

$$\mathbf{x}^{-1} = |\mathbf{x}|^{-1}\, e^{-i\arg\mathbf{x}} \tag{11.13}$$

and in rectangular form it is

$$\mathbf{x}^{-1} = |\mathbf{x}|^{-2}\,\mathbf{x}^* \tag{11.14}$$

where $\mathbf{x}^* = \operatorname{Re}\mathbf{x} - i\operatorname{Im}\mathbf{x}$ is the complex conjugate of **x**. Division is multiplication by the reciprocal, so that

$$\mathbf{x}/\mathbf{y} = \mathbf{xy}^{-1} = |\mathbf{x}||\mathbf{y}|^{-1}\, e^{i(\arg\mathbf{x} - \arg\mathbf{y})} \tag{11.15}$$

We note that $\mathbf{xx}^* = |\mathbf{x}|^2 \neq \mathbf{x}^2$.

The natural logarithm of **x** is defined from

$$\mathbf{x} = |\mathbf{x}|e^{i\arg\mathbf{x}} \tag{11.16}$$

by taking the logarithm of both sides, as

$$\ln\mathbf{x} = ln|\mathbf{x}| + i\arg\mathbf{x} \tag{11.17}$$

This is a multiple-valued function.[5] The principal value of ln **x** is defined from

[5] Logarithms of positive real numbers are also multiple valued. For example $ln\ e = 1$; but also $ln\ e = 1 \pm i2\pi,\ 1 \pm i4\pi$, etc.

$$\mathbf{x} = |\mathbf{x}|e^{i \operatorname{Arg} \mathbf{x}} \tag{11.18}$$

as

$$\operatorname{Ln} \mathbf{x} = \operatorname{Ln}|\mathbf{x}| + i \operatorname{Arg} \mathbf{x} \tag{11.19}$$

Equations (11.17) and (11.19) do not tell us how to find the ordinary logarithm (the logarithm for a real, positive argument); however, they do tell us how to find the "extraordinary" logarithm (the logarithm for a negative or complex argument) in terms of the ordinary logarithm. We examine this "organic" form further in a subsequent section.

The signum function of **x** is defined as

$$\operatorname{sgn} \mathbf{x} = \mathbf{x}/\,|\mathbf{x}| = e^{i \arg \mathbf{x}} = e^{i \operatorname{Arg} \mathbf{x}} \tag{11.20}$$

It follows that

$$|\operatorname{sgn} \mathbf{x}| = 1, \tag{11.21}$$

$$(\operatorname{sgn} \mathbf{x})^{-1} = e^{-i \arg \mathbf{x}} = (\operatorname{sgn} \mathbf{x})^* = \operatorname{sgn}(\mathbf{x}^*) \tag{11.22}$$

From Eq. (11.20) we obtain

$$\arg \mathbf{x} = -i \ln \operatorname{sgn} \mathbf{x} \tag{11.23a}$$

$$\operatorname{Arg} \mathbf{x} = -i \operatorname{Ln} \operatorname{sgn} \mathbf{x} \tag{11.23b}$$

which shows that *Arg* is a kind of "imaginary logarithm." One might call Arg **x** the *Argandian logarithm* of **x**.

The exponential $e^{i\theta}$ can be considered to be a generalization of the concept *sign,* because it is +1 when $\theta = 0$, i when $\theta = \frac{1}{2}\pi$, -1 when $\theta = \pi$, $-i$ when $\theta = \frac{3}{2}\pi$, +1 again when $\theta = 2\pi$, etc.[6] Thus it is appropriate that the signum function of a complex number equals $e^{i\theta}$ because *signum* is Latin for sign.

11.2 Electronic Representation of a Complex Variable

To fully specify a complex variable **x** we must specify two real variables. Two choices are either Re **x** and Im **x**, or $|\mathbf{x}|$ and Arg **x**. Thus one way to represent a complex variable electronically is by the use of two signal paths. For example, to add complex voltages **u** and **v** we perform the operations

$$\operatorname{Re}(\mathbf{u} + \mathbf{v}) = \operatorname{Re} \mathbf{u} + \operatorname{Re} \mathbf{v} \tag{11.24a}$$

$$\operatorname{Im}(\mathbf{u} + \mathbf{v}) = \operatorname{Im} \mathbf{u} + \operatorname{Im} \mathbf{v} \tag{11.24b}$$

and to multiply we perform the operations[7]

$$|\mathbf{uv}| = |\mathbf{u}||\mathbf{v}| \tag{11.25a}$$

$$\measuredangle (\mathbf{uv}) = \operatorname{Arg} \mathbf{u} + \operatorname{Arg} \mathbf{v} \tag{11.25b}$$

[6]This concept of sign is valid for any θ-value between $-\infty$ and $+\infty$, not just the values used in the examples here. Thus in general $e^{i\theta} = x + ix'$ (with $|x| = 1$) can legitimately be called a sign.

[7]In Eq. (11.25b) $\measuredangle (\mathbf{uv})$ is single valued, but $\measuredangle (\mathbf{uv}) \neq \operatorname{Arg} (\mathbf{uv})$. Instead, $\operatorname{saw}_{2\pi} \measuredangle (\mathbf{uv}) = \operatorname{Arg} (\mathbf{uv})$.

In block diagram form these relationships appear as shown in Figs. 11.2 and 11.3, respectively.

In the first case (addition) we work with the complex variables in terms of rectangular coordinates, whereas in the second case it is more convenient to work in terms of polar coordinates. This points up the need for circuits that convert from one set of coordinates to the other. Such circuits are called *coordinate converters*. (Sometimes the term *resolver* is used but this term is more often used to refer to a particular electromechanical device.)

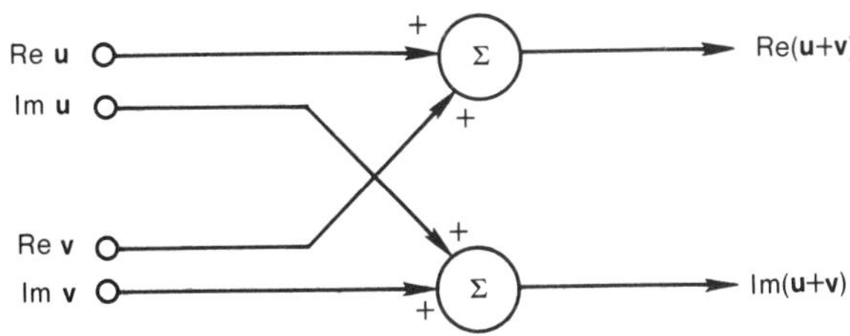

Figure 11.2 Block diagram of the addition of two complex signals.

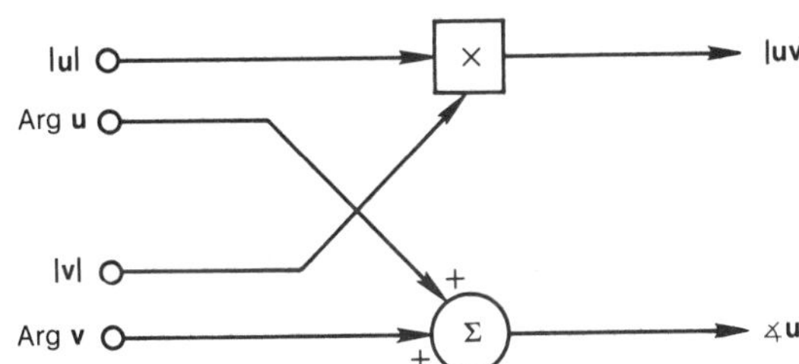

Figure 11.3 Block diagram of the multiplication of two complex signals.

A *polar-to-rectangular converter* performs the operations

$$\text{Re}\,\mathbf{u} = |\mathbf{u}| \cos \text{Arg}\,\mathbf{u} \tag{11.26a}$$

$$\text{Im}\,\mathbf{u} = |\mathbf{u}| \sin \text{Arg}\,\mathbf{u} \tag{11.26b}$$

The block diagram for this case is shown in Fig. 11.4.

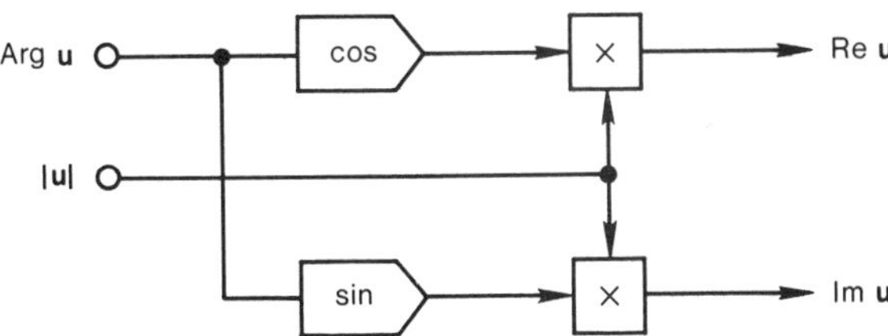

Figure 11.4 Block diagram of polar to rectangular converter.

A *rectangular-to-polar converter* performs the operations

$$|\mathbf{u}| = [(\text{Re}\,\mathbf{u})^2 + (\text{Im}\,\mathbf{u})^2]^{1/2} \tag{11.27a}$$

$$\text{Arg}\,\mathbf{u} = \Theta \tag{11.27b}$$

where Θ can be as defined in Eqs. (11.6), (11.7), or (11.8), or it can be simply $\text{Tan}^{-1}(\text{Im}\,\mathbf{u}\ /\ \text{Re}\,\mathbf{u})$ if the range of values for Θ between $-\frac{1}{2}\pi$ and $\frac{1}{2}\pi$ is adequate. A method appearing in Korn and Korn[8] can be used to solve for Θ implicitly as follows. First we note that $x' \cos\Theta$ and $x \sin\Theta$ both equal xx'/r, so that

$$\text{Im}\,\mathbf{u}\cos\Theta - \text{Re}\,\mathbf{u}\sin\Theta = 0 \tag{11.28}$$

This equation is represented by the block diagram of Fig. 11.5, providing a solution for Θ that does not require any division.

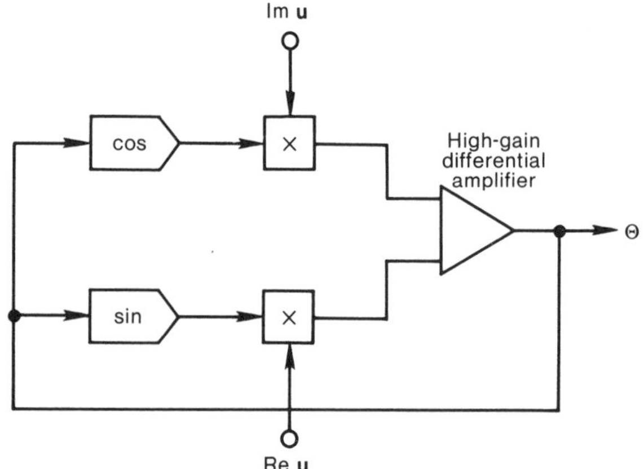

Figure 11.5 Block diagram of the implicit solution of Eq. (11.28) for Θ.

An implicit solution is also possible for finding $|\mathbf{u}|$ using the equation

$$|\mathbf{u}| = \frac{(\text{Re}\,\mathbf{u})^2}{|\mathbf{u}| + \text{Im}\,\mathbf{u}} + \text{Im}\,\mathbf{u} \tag{11.29}$$

as given in Sheingold (op.cit., page 21). Sheingold reports that circuits based on this relationship are more accurate than those based on Eq. (11.27a).

Another way to represent a complex variable electronically is as the amplitude and phase of a sinusoid. The amplitude is used to carry information of $|\mathbf{u}|$ and the phase to carry information of Arg $\mathbf{u}$. This method is used to carry the color information (chrominance) on the color subcarrier in

[8]G. A. Korn and T. M. Korn, *Electronic Analog Computers*, 2nd ed. (New York: McGraw-Hill, 1956) p. 336.

the American NTSC (National Television System Committee) color television system. The reader is referred to works on color television for a description of this system.[9]

This latter method has the advantage of requiring only one signal path; however, signal frequencies must be much lower than the carrier frequency. The previous method with two signal paths can also be multiplexed onto one signal path—at least in principle.

11.3 The Complete Logarithmic Function

Logarithmic amplifiers are useful for compressing the dynamic range of signals. Mathematically, the logarithm tends to negative infinity as its argument tends to zero from more positive values. Also, the logarithm is not real (in the mathematical sense) for negative values of the argument. How are these facts manifested in the case of logarithmic amplifiers? How can we "work around" them? What, really, is a logarithmic amplifier? One purpose of this section is to present answers to these questions.

One might reasonably expect that functions of a complex variable would be complex. On the other hand, functions of a real variable can also have imaginary or complex values. Such cases occur for negative values of the (real) argument for some functions. Examples are $\mathrm{Ln}\, x$ and $x^{1/2}$. Cases also occur for nonnegative values of the argument for some functions. An example is $\mathrm{Sin}^{-1}x$, which is real only for $-1 \leqslant x \leqslant 1$.

In this and the next section we show how the traditional definitions of some of these functions can be extended from the region where they are real into the region where they are complex. We also show how such functions can be visualized.

For convenience, the following terminology is introduced. The portion of a function that is real is called the *ordinary part;* the portion that is complex (or at least imaginary) is called the *extraordinary part;* and the combination of both parts is called the *complete function.*

We consider first the complete logarithmic function, $f(x) = \mathrm{Ln}\, x$. For negative values of x, $\mathrm{Ln}\, x$ is customarily defined as

$$\mathrm{Ln}\, x = \mathrm{Ln}(-x) + i\pi, \; x < 0 \qquad (11.30)$$

[9]The definitive works on this subject are: (1) *Proceedings of the IRE,* ed. Alfred N. Goldsmith, vol. 39, no. 10, (New York: Institute of Electrical and Electronics Engineers, October 1951); (2) *Proceedings of the IRE,* ed. Alfred N. Goldsmith, vol. 42, no. 1, (New York: Institute of Electrical and Electronics Engineers, January 1954).

so that for all x-values we can write

$$\operatorname{Ln} x = \operatorname{Ln}|x| + i\pi S(-x) \tag{11.31}$$

Equation (11.30) gives the extraordinary part, Eq. (11.31) gives the complete function, and the ordinary part of Ln x exists for $x > 0$. (The "point" at $-\infty$, corresponding to $x = 0$, is uncategorized at this time as to whether it is ordinary or extraordinary.)

The right-hand side of Eq. (11.31) is the *organic form* of the function Ln x. Equation (11.31) has the following properties that are characteristic of organic forms in general:

1. It pretends to define the complete function.
2. It does indeed define the extraordinary part in terms of the ordinary part.
3. It reduces to the null equation $0 = 0$ [or $f(x) = f(x)$] for the ordinary part, *except that it may truly define isolated points of the ordinary part.*

A function need not have an extraordinary part in order for it to have an organic form. Examples are $\operatorname{Tan}^{-1}x$ and $\operatorname{Ctn}^{-1}x$, considered later.

A preliminary definition of "organic form" is *the expression obtained by writing a function in terms of complex quantities.* Thus, although $\operatorname{Tan}^{-1}x$ is nowhere imaginary or complex, it can nevertheless be expressed in terms of complex quantities. This is its organic form. Cf. Table 11.1.

The traditional way to visualize these functions is to graph only the ordinary part. For the general case this can be written as the two-dimensional shaded spaceform

$$\left.\begin{aligned} x &= x \\ y &= \operatorname{Re} f(x) \\ z &= \delta^\circ[\operatorname{Im} f(x)] \end{aligned}\right\} \tag{11.32}$$

When $\operatorname{Im} f(x) \neq 0$, $z = 0$. Thus the extraordinary part is blanked. For the case $f(x) = \operatorname{Ln} x$, this spaceform is

$$\left.\begin{aligned} x &= x \\ y &= \operatorname{Ln}|x| \\ z &= S(x) \end{aligned}\right\} \tag{11.33}$$

where we have made use of the relationship

$$\delta^\circ[\pi S(-x)] = \delta^\circ[S(-x)] = S(x) \tag{11.34}$$

Another way to visualize these functions is to plot the real part and the coefficient of the imaginary part of the organic form against the independent variable. This gives in general

TABLE 11.1 Some Organic Forms

The organic forms for the six inverse trigonometric functions are all of the form $-i$ Ln u, where $u = u(x)$ is as follows:

Function	u
$\mathrm{Sin}^{-1}x$	$(1 - x^2)^{1/2} + ix$
$\mathrm{Cos}^{-1}x$	$x + i(1 - x^2)^{1/2}$
$\mathrm{Tan}^{-1}x$	$(1 + ix)/(1 + x^2)^{1/2}$
$\mathrm{Ctn}^{-1}x$	$(x + i)/(1 + x^2)^{1/2}$
$\mathrm{Sec}^{-1}x$	$1/x + i(x^2 - 1)^{1/2}/x$
$\mathrm{Csc}^{-1}x$	$(x^2 - 1)^{1/2}/x + i/x$

$$\left.\begin{aligned} x &= x \\ y &= \mathrm{Re}\,f(x) \\ y' &= \mathrm{Im}\,f(x) \end{aligned}\right\} \tag{11.35}$$

This is a three-dimensional spaceform of the complete function. In the present case this spaceform is

$$\left.\begin{aligned} x &= x \\ y &= \mathrm{Ln}|x| \\ y' &= \pi S(-x) \end{aligned}\right\} \tag{11.36}$$

A representation of the complete logarithmic function of a real variable is presented in Fig. 11.6. When the two indicated folds are made, this becomes a graph of Ln x as defined in Eq. (11.36) in three dimensions. This *origami* (paper folding) method of graphing is useful for three-dimensional curves that are developable onto a plane surface. These include curves that consist entirely of two-dimensional branches as Ln x does.

We note that the traditional visualization of this function, as defined by Eqs. (11.33), is just a section taken through the complete spaceform in the $y' = 0$ plane.

The block diagram of the complete logarithmic function as defined by Eq. (11.31) is shown in Fig. 11.7. Circuits based on this block diagram are usable for both positive and negative x-values. However, they are not usable for values of x at or near zero because $\mathrm{Ln}|x|$ becomes large (in the negative direction) without limit as $x \to 0$, and real electronic circuits have a finite dynamic range. For this reason, the function $\sinh^{-1}x$

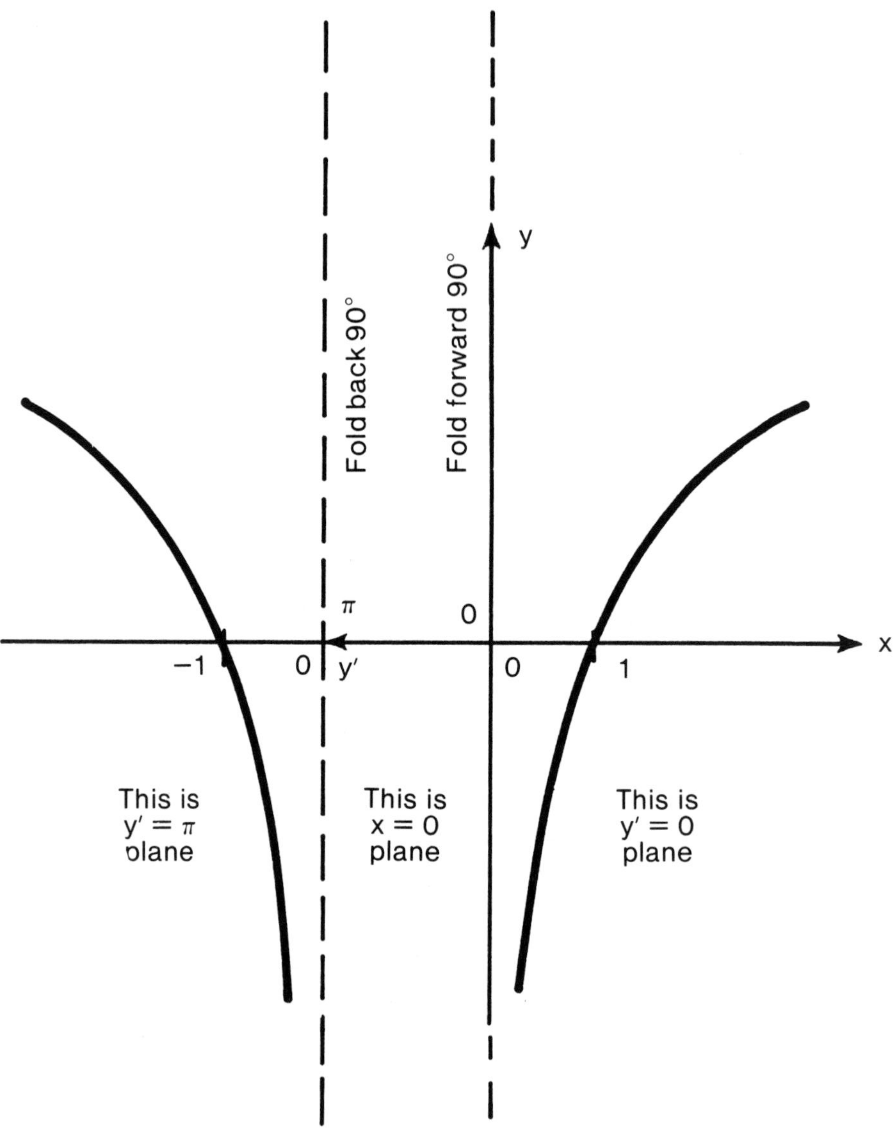

Figure 11.6 The complete logarithmic function of a real variable.

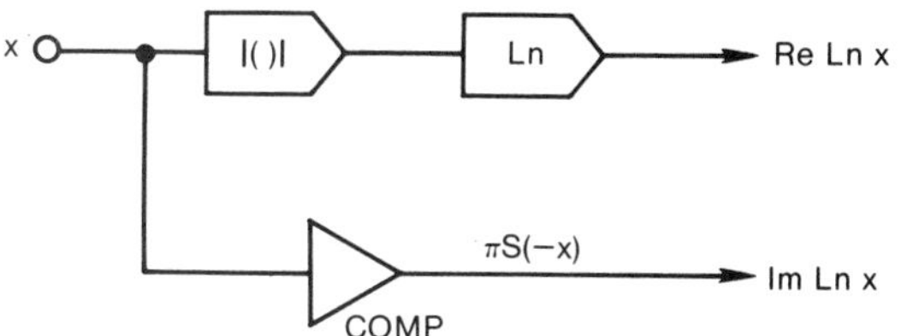

Figure 11.7 Block diagram of the complete logarithmic function of a real variable.

is sometimes substituted for Ln x. This function is asymptotic to sgn x Ln$|2x|$ as was pointed out in Chapter 8. This so-called "a-c log" can be used to compress the dynamic range of signals (one of the uses for logarithmic circuits), while having the advantage of being usable with bipolar inputs that go through zero. However, if a truly logarithmic function is required, then either the input must be biased so that it does not go through zero, or a range of x-values around zero must be excluded. So-called *bipolar logarithmic amplifiers* generally use an approximation of the $\sinh^{-1}$ function.[10]

11.4 The Complete Inverse Sine Function

Our investigation of complex functions continues with the inverse sine function. One purpose of examining the complete inverse sine function is to show why the principal value of $\sin^{-1}x$ is defined as it is in Table 1.1. Another purpose is to permit visualization of the three-dimensional spaceform of that function. A third purpose is to find its block diagram.

The organic form of the function $\text{Sin}^{-1}x$ can be obtained by consideration of the complex quantity $\mathbf{u}$ where $|\mathbf{u}| = 1$ and Arg $\mathbf{u} = \Theta = \text{Sin}^{-1}x$. This complex quantity is diagrammed in Fig. 11.8. We can write $\mathbf{u}$ in two ways, which we equate here:

$$\mathbf{u} = e^{i\Theta} = (1 - x^2)^{1/2} + ix \qquad (11.37)$$

Solving this equation for $\Theta = \text{Sin}^{-1}x$ gives the organic form for this case:

$$\text{Sin}^{-1}x = -\text{i}\,\text{Ln}[(1 - x^2)^{1/2} + ix] \qquad (11.38)$$

$$\text{Sin}^{-1}x = -\text{i}\,\text{Ln}[i(x^2 - 1)^{1/2} + ix] \qquad (11.39)$$

The extraordinary part of this exists for $|x| > 1$ and can be found from Eq. (11.39) as

$$\text{Sin}^{-1}x = \tfrac{1}{2}\pi \operatorname{sgn} x - i\,\text{Ln}|(x^2 - 1)^{1/2} + x|, \quad |x| > 1 \qquad (11.40)$$

[10] D. H. Sheingold, ed., *Nonlinear Circuits Handbook,* 2nd ed., by the Engineering Staff of Analog Devices, Inc. (Norwood, Mass.: Analog Devices, Inc., January 1976). See page 102 for example.

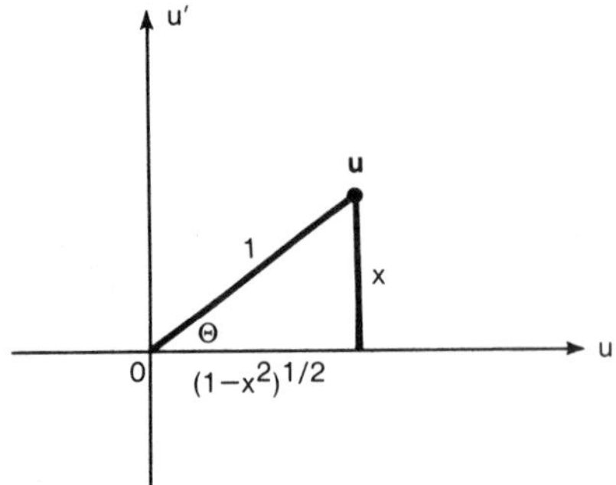

Figure 11.8 Argand diagram for finding the organic form of $\mathrm{Sin}^{-1}x$.

Finally, the organic form of $\mathrm{Sin}^{-1}x$ can be separated into its real part and coefficient of its imaginary part for all x-values as

$$\mathrm{Re\ Sin}^{-1}x = \mathrm{Sin}^{-1}(\mathrm{clip}^{1}_{-1}x) \tag{11.41a}$$

$$\mathrm{Im\ Sin}^{-1}x = -u\ \mathrm{Ln}|(x^2 - 1)^{1/2} + x|, \quad u = 1 - \mathrm{puls}^{1}_{-1}x \tag{11.41b}$$

In the expression for Im $\mathrm{Sin}^{-1}x$, we can replace $(x^2 - 1)$ with $|x^2 - 1|$ because when $(x^2 - 1)$ is negative, u is zero, so that Im $\mathrm{Sin}^{-1}x = 0$ there in either case. Therefore,

$$\mathrm{Im\ Sin}^{-1}x = -u\ \mathrm{Ln}||x^2 - 1|^{1/2} + x| \tag{11.42}$$

The two functions Re $\mathrm{Sin}^{-1}x$ and Im $\mathrm{Sin}^{-1}x$ are graphed in Fig. 11.9.

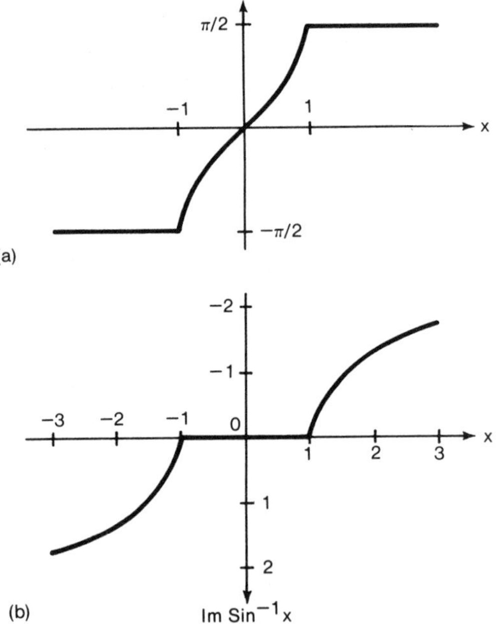

Figure 11.9 The complete inverse sine function: (a) Re $\mathrm{Sin}^{-1}x$; (b) Im $\mathrm{Sin}^{-1}x$.

The traditional visualization of $\mathrm{Sin}^{-1}x$ is represented by the two-dimensional shaded spaceform

$$\left.\begin{aligned} x &= x \\ y &= \mathrm{Re}\,\mathrm{Sin}^{-1}x \\ z &= \mathrm{puls}_{-1}^{1}x \end{aligned}\right\} \qquad (11.43)$$

where use has been made of the relationship

$$\begin{aligned} \delta^{\circ}(\mathrm{Im}\,\mathrm{Sin}^{-1}x) &= \delta^{\circ}(-u) \\ &= \delta^{\circ}(\mathrm{puls}_{-1}^{1}x - 1) \\ &= \mathrm{puls}_{-1}^{1}x \end{aligned} \qquad (11.44)$$

The complete visualization of $\mathrm{Sin}^{-1}x$ is represented by the three-dimensional spaceform

$$\left.\begin{aligned} x &= x \\ y &= \mathrm{Re}\,\mathrm{Sin}^{-1}x \\ y' &= \mathrm{Im}\,\mathrm{Sin}^{-1}x \end{aligned}\right\} \qquad (11.45)$$

A three-dimensional graph of this spaceform can be made from Fig. 11.10 by making the two indicated folds. From this graph, the complete $\mathrm{Sin}^{-1}x$ function is seen to be a three-dimensional continuous-broken function.

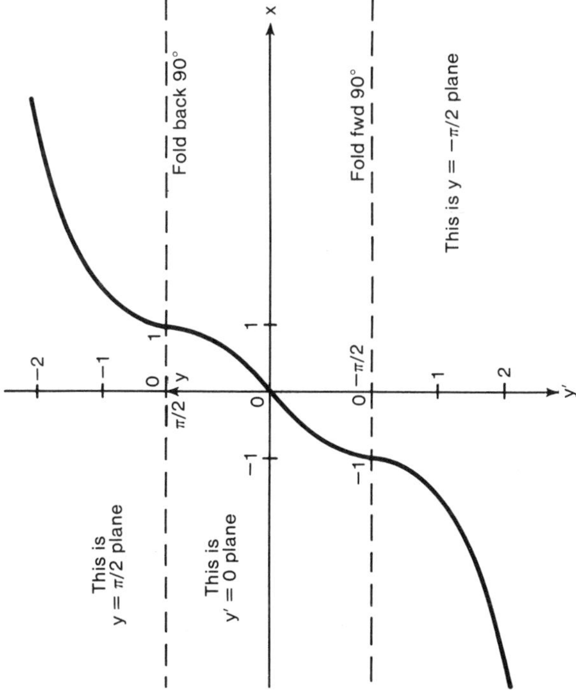

Figure 11.10 The complete inverse sine function.

A block diagram of $\mathrm{Sin}^{-1}x$ based on Eqs. (11.41a) and (11.42) is shown in Fig. 11.11. This block diagram is good for all x-values. Since the argument of Ln is always positive (and never zero), there is no difficulty in taking the logarithm over the full range of x-values of both signs.

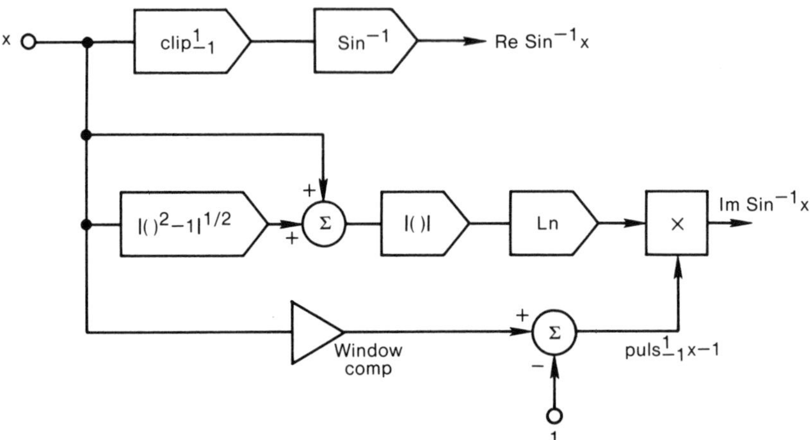

Figure 11.11 Block diagram of the complete inverse sine function.

The organic form for $\mathrm{Sin}^{-1}x$ in Eq. (11.38) can be differentiated to obtain the classical derivative of $\mathrm{Sin}^{-1}x$, namely

$$\frac{d}{dx}\mathrm{Sin}^{-1}x = (1 - x^2)^{-1/2} \tag{11.46}$$

This expression holds for the extraordinary part as well as for the ordinary part.

The organic forms for the remaining five inverse trigonometric functions can be found in a manner similar to that used above in the case of $\mathrm{Sin}^{-1}x$. They are listed in Table 11.1, along with $\mathrm{Sin}^{-1}x$. The principal branches of those six functions can be found from their organic forms as a result of Property 3 of organic forms. For example, it is easy to show that $\mathrm{Sin}^{-1}x = 0$ when $x = 0$, and that it has the value $\frac{1}{2}\pi$ when $x = 1$ and the value $-\frac{1}{2}\pi$ when $x = -1$. Inverse cosine, tangent, and cotangent functions can similarly be confirmed to correspond to the definitions given in Table 1.1 of Chapter 1. Inverse secant and cosecant functions present a special problem that is discussed next.

First, we note that the customary definition of Ln x given in Eq. (11.31) is somewhat arbitrary in view of the fact that the definition of Ln $\mathbf{x}$, with $x' = 0$, would allow it to be either of the following:

$$\text{Ln}\, x = \text{Ln}|x| + i\pi \text{S}(-x) \tag{11.47a}$$

$$\text{Ln}\, x = \text{Ln}|x| - i\pi \text{S}(-x) \tag{11.47b}$$

In most applications we encounter no problems if we arbitrarily choose it to be the first of these.

Actually, the choice of sign for the imaginary part of Ln x cannot be made arbitrarily; for Ln x must be defined consistently with the defining expression for Ln $\mathbf{x}$ in Eq. (11.19). In that equation, Ln $\mathbf{x}$ is seen to exhibit infinitesimal hysteresis wherein the plus sign for the imaginary part of Ln x is indicated for the positive direction of progression of Θ and the minus sign for the negative direction. Making use of this fact, it is not difficult to show that the principal branches of the inverse secant and cosecant functions are as shown in Table 1.1 and Fig. 1.7. The case $\text{Sec}^{-1}x$ is worked out next.

The organic form of $\text{Sec}^{-1}x$ is

$$\text{Sec}^{-1}x = -i\, \text{Ln} \frac{1 + i(x^2 - 1)^{1/2}}{x} \tag{11.48}$$

giving, for the positive branch of the principal value, the "points"[11]

$$\text{Sec}^{-1}(1) = -i\, \text{Ln}(1) = 0 \tag{11.49a}$$

$$\text{Sec}^{-1}(\infty) = -i\, \text{Ln}(i) = -i(i\tfrac{1}{2}\pi) = \tfrac{1}{2}\pi, \tag{11.49b}$$

and for the negative branch

$$\text{Sec}^{-1}(-1) = -i\, \text{Ln}(-1) = -i(-i\pi) = -\pi \tag{11.50a}$$

$$\text{Sec}^{-1}(-\infty) = -i\, \text{Ln}(-\text{i}) = -i(-i\tfrac{1}{2}\pi) = -\tfrac{1}{2}\pi \tag{11.50b}$$

Note that Ln(-1) in Eq. (11.50a) is given as $-i\pi$ where normally it would be taken to be $i\pi$. This is due to the fact that this point is approached from the "point" at $(-\infty, -\frac{1}{2}\pi)$ by letting x become more positive, thus making $\text{Sec}^{-1}x = \Theta$ progress in the negative direction. This takes us to the point at $(-1, -\pi)$ because of infinitesimal hysteresis. Refer to Fig. 1.7.

11.5 Visualizing the Four-Dimensional Logarithmic Function

Having come this far, we would be remiss if we did not examine the complete logarithm where the independent variable is complex. The principal value of the natural logarithm for this case is defined as

$$\text{Ln}\, \mathbf{x} = \text{Ln}|\mathbf{x}| + i\, \text{Arg}\, \mathbf{x} \tag{11.51}$$

[11] $\text{Sec}^{-1}(\infty)$ is a shorthand notation meaning $\lim\limits_{x\to\infty} \text{Sec}^{-1}x$.

In general we define the spaceform representing the function of a complex variable $\mathbf{y} = f(\mathbf{x})$ as

$$\left.\begin{aligned} x &= \operatorname{Re}\mathbf{x} \\ x' &= \operatorname{Im}\mathbf{x} \\ y &= \operatorname{Re}\mathbf{y} \\ y' &= \operatorname{Im}\mathbf{y} \end{aligned}\right\} \quad (11.52)$$

This is a four-dimensional spaceform. The quantities x and x' are both independent variables; y and y' are dependent variables. Therefore the spaceform representation of Ln $\mathbf{x}$ is

$$\left.\begin{aligned} x &= x \\ x' &= x' \\ y &= \operatorname{Ln}|\mathbf{x}| \\ y' &= \operatorname{Arg}\mathbf{x} \end{aligned}\right\} \quad (11.53)$$

For visualization purposes, the four-dimensional spaceform of Eqs. (11.53) can be completely represented as the following three-dimensional shaded spaceform:

$$\left.\begin{aligned} x &= x \\ x' &= x' \\ y &= \operatorname{Ln}|\mathbf{x}| \\ z &= \tfrac{1}{2}(\operatorname{Arg}\mathbf{x})/\pi + \tfrac{1}{2} \end{aligned}\right\} \quad (11.54)$$

Since Arg $\mathbf{x}$ only assumes values between $-\pi$ and π, then z only assumes values between zero and one. Thus we have coded all of y' into all of z. In this way the observer can gain an appreciation for the *Gestalt* of the entire spaceform without actually being required to view a four-dimensional visual object.

The shaded spaceform of Eqs. (11.54) is an infinitely-long curving funnel-shaped surface that varies linearly in intensity as Θ goes from $-\pi$ to π. The y-axis is the funnel's axis of rotational symmetry.

A block diagram of Ln $\mathbf{x}$ is shown in Fig. 11.12. It is important to note that the two inputs to the block diagram are in polar form whereas the outputs are in rectangular form.

11.6 Fractional Exponents

One application of analog circuits is to the generation of arbitrary powers of an input signal. This is accomplished by use of a device (module or integrated circuit) called a *multifunction converter*.

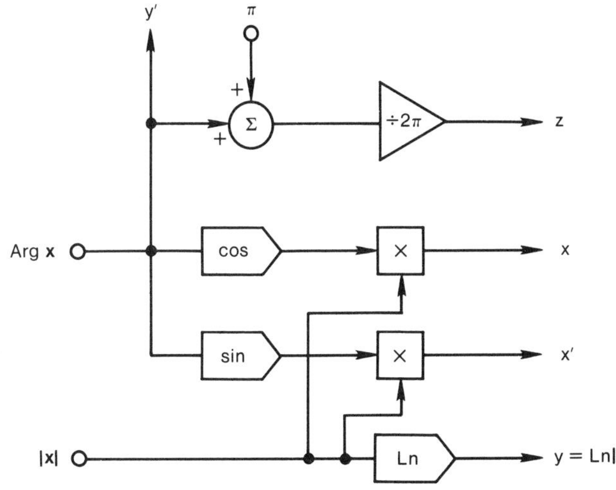

Figure 11.12 Block diagram of the complete logarithmic function of a complex variable.

This section discusses the function $y = x^a$, where a is any real number. In general, x^a is complex. The ramifications of this fact on the function, its derivative, its integral, and its electronic representation are discussed.

De Moivre's theorem[12] provides a method for finding the multiple values of roots: two values for the square root (1/2 power), three for the cube root (1/3 power), and so on. Indeed, any rational fractional power can be found by use of that theorem.

In what follows, we find instead the *principal value* of fractional powers, whether they be *rational* or *irrational* fractional powers. One reason these are important is that integrals and derivatives dealing with fractional exponents assume principal values. Another reason is that electronic devices that generate such functions generate single-valued functions of necessity. Such devices are called *multifunction converters*.[13] We treat only the case of real x-values.

[12] See virtually any text on algebra or trigonometry for a review of this important theorem.

[13] See for example pages 113 ff. of Y. J. Wong & W. E. Ott, *Function Circuits, Design and Applications,* McGraw-Hill, New York, 1976.

In Chapter 1, $(\ )^{1/n}$ was defined as the principal value of the nth root. The organic form of x^a can be found from

$$x^a = |x|^a(\text{sgn}\, x)^a \tag{11.55}$$

For nonzero x-values, the first factor is defined as

$$|x|^a = \exp(a\, \text{Ln}|x|),\ x \neq 0 \tag{11.56}$$

with exp(0) being defined as 1, and $|0|^a$ being defined as 0 except when $a \leqslant 0$. The case $|0|^\circ$ is undefined.

For negative x-values, Eq. (11.55) becomes

$$x^a = (-x)^a(-1)^a,\ x < 0 \tag{11.57}$$

Since -1 can be written $e^{i\pi}$, we have

$$(-1)^a = e^{i\pi a} \tag{11.58}$$

This function represents the locus of points on the unit circle centered on the origin in the complex plane, with $\theta = \pi a$.

Finally, we have, for all real x-values,[14]

$$(\text{sgn}\, x)^a = S(x) + e^{i\pi a}S(-x) \tag{11.59}$$

and

$$x^a = |x|^a[S(x) + e^{i\pi a}S(-x)] \tag{11.60}$$

which is the organic form of x^a.

In general one can write

$$\mathbf{y} = x^a = y + iy' \tag{11.61}$$

where y and y' are given for specific a-values in the following examples.

Example 11.1. Let $a = 1$. $x^a = |x|[S(x) - S(-x)]$;
$y = |x|\text{sgn}\, x = x,\ y' = 0.$

Example 11.2. Let $a = 1.5$. $x^a = |x|^{1.5}[S(x) - iS(-x)]$;
$y = |x|^{1.5}S(x),\ y' = -|x|^{1.5}S(-x).$

[14] It can also be shown that $(\text{sgn}\, x)^a = \text{sgn}\, \mathbf{u}$ where $\mathbf{u} = x^a$. Proof follows:

1. $(\mathbf{x})^a = (|\mathbf{x}|\ e^{i\,\text{Arg}\,\mathbf{x}})^a = |\mathbf{x}|^a\ e^{ia\,\text{Arg}\,\mathbf{x}}$
2. $(\mathbf{x}^a) = |\mathbf{x}^a|\ e^{i\,\text{Arg}(\mathbf{x}a)}$
3. But $(\mathbf{x}^a) = (\mathbf{x})^a = \mathbf{x}^a$; therefore,
$$|\mathbf{x}^a|\ e^{i\,\text{Arg}(\mathbf{x}^a)} = |\mathbf{x}|^a\ e^{ia\,\text{Arg}\,\mathbf{x}}.$$
4. If two complex quantities are equal, this means their magnitudes are equal. Thus,
$$|\mathbf{x}^a| = |\mathbf{x}|^a.$$
5. Further,
$$\text{sgn}(\mathbf{x}^a) = \frac{\mathbf{x}^a}{|\mathbf{x}^a|} = \frac{\mathbf{x}^a}{|\mathbf{x}|^a} = \left(\frac{\mathbf{x}}{|\mathbf{x}|}\right)^a = (\text{sgn}\ \mathbf{x})^a.$$
6. Finally, let $\mathbf{x} \to x$.

Example 11.3. Let $a = 2$. $x^a = |x|^2[S(x) + S(-x)]$;
$y = |x|^2 = x^2$, $y' = 0$.

Example 11.4. Let $a = 2.5$. $x^a = |x|^{2.5}[S(x) + iS(-x)]$;
$y = |x|^{2.5}S(x)$, $y' = |x|^{2.5}S(-x)$.

Example 11.5. Let $a = 3$. $x^a = |x|^3[S(x) - S(-x)]$;
$y = |x|^3 \text{sgn}\, x = x^3$, $y' = 0$.

Comparison of these examples in sequence shows the following. In the three-dimensional space defined by (x, y, y'), x^a for negative x-values rotates around the x-axis as a varies monotonically. Thus when $a = 1$, **y** is negative (for negative x-values); when $a = 1.5$, **y** is negative imaginary; when $a = 2$, **y** is positive; when $a = 2.5$, **y** is positive imaginary; and when $a = 3$, **y** is negative again.

The above examples all use positive a-values, but negative a-values present no problem because $\exp(-u) = 1/\exp u$ so that $|x|^{-a} = 1/|x|^a$. For example,

$$\mathrm{x}^{-1/2} = (1/|x|^{1/2})[S(x) - iS(-x)] \tag{11.62}$$

Equation (11.60) can be differentiated (a = const.) to obtain

$$\frac{d}{dx}x^a = ax^{a-1} + (1 - e^{i\pi a})\,\delta^{-a+1}(x) \tag{11.63}$$

where $\delta^{-a+1}(x)$ means $|x|^a\delta(x)$.

Examples of the use of this equation follow.

Example 11.6. Let $a = -m$ where m is a positive odd integer.

$$\frac{d}{dx}x^{-m} = -mx^{-m-1} + 2\delta^{m+1}(x)$$

This result compares favorably with that from Example 9.1 in Chapter 9.

Example 11.7. Let $a = n$, an even integer.

$$\frac{d}{dx}x^n = nx^{n-1}$$

because $(1 - e^{i\pi n}) = (1 - 1) = 0$.

Example 11.8. Let $a > 0$.

$$\frac{d}{dx}x^a = ax^{a-1}$$

because $\delta^{-a+1}(x)$ is a null function.

The integral of x^b can be found by integrating Eq. (11.63) and substituting b for $a - 1$. It is

$$\int x^b dx = \frac{1}{b+1}x^{b+1} - \frac{1}{b+1}[1 - e^{i\pi(b+1)}]\,\delta^{-b-1}(0)S(x) + C,\ \mathrm{b} \neq -1 \qquad (11.64)$$

Examples of the use of this equation follow.

Example 11.9. Let $b = n$, an even integer.

$$\int x^n dx = \frac{1}{n+1}x^{n+1} - \frac{2}{n+1}\,\delta^{-n-1}(0)S(x) + C$$

Example 11.10. Let $b = m$, an odd integer.

$$\int x^m dx = \frac{1}{m+1}x^{m+1} + C,\ m \neq -1$$

Example 11.11. Let $b > -1$.

$$\int x^b dx = \frac{1}{b+1}x^{b+1} + C,\ b > -1$$

Example 11.12. Find the integral of x^b from zero to x.

$$\int_0^x x^b dx = \frac{1}{b+1}x^{b+1},\ b \neq -1$$

The second term containing $S(x)$ drops out because $S(x) = S(0)$ for the following reasons: If the interval of integration is to positive x, then $S(x) = S(0) = 1$; and if it is to negative x, then $S(x) = S(0) = 0$ because of infinitesimal hysteresis.

We have seen that x^a is sometimes real, sometimes imaginary, and in general complex. Sometimes it has a unique inverse, but in general it does not. These complications do not arise if x is restricted to positive values.

In the electronic applications one can often work in terms of $|x|^a$ or $|x|^a\mathrm{sgn}\,x$ instead of x^a. In these cases, $|x|^a = x^a$ when a is an even integer, and $|x|^a\mathrm{sgn}\,x = x^a$ when a is an odd integer. Of course $|x|^a$ is always an even function and $|x|^a\mathrm{sgn}\,x$ is always an odd function for any real a-value.

In the following paragraph we examine the derivatives and integrals of $|x|^a$ and $|x|^a\mathrm{sgn}\,x$.

The derivatives are

$$\frac{d}{dx}|x|^a = a|x|^{a-1}\mathrm{sgn}\,x \qquad (11.65)$$

$$\frac{d}{dx}|x|^a\mathrm{sgn}\,x = a|x|^{a-1} + 2\,\delta^{-a+1}(x),\ -\infty < a < \infty \qquad (11.66)$$

$$= a|x|^{a-1}, a > 0 \tag{11.67}$$

From those expressions one can easily obtain the integrals. The integral of $|x|^b$ is

$$\int |x|^b\, dx = \frac{1}{b+1}|x|^{b+1}\text{sgn } x - \frac{2}{b+1}\,\delta^{-b-1}(0)S(x) + C, \qquad b \neq -1 \tag{11.68}$$

For $b > -1$ this becomes

$$\int |x|^b\, dx = \frac{1}{b+1}|x|^{b+1}\text{sgn } x + C \tag{11.69}$$

The definite integral from zero to x is

$$\int_0^x |x|^b\, dx = \frac{1}{b+1}|x|^{b+1}\text{sgn } x, \qquad b \neq -1 \tag{11.70}$$

The integral of $|x|^b\text{sgn } x$ is

$$\int |x|^b \text{sgn } x\, dx = \frac{1}{b+1}|x|^{b+1} + C, \qquad b \neq -1 \tag{11.71}$$

Another useful alternative to x^a is the *invertible power, Inv* x^a, defined as follows.[15]

When we allow the existence of complex numbers, the square root of a real number is *invertible*. That is, by applying the inverse operation (squaring in this case), we obtain the original quantity, so that $(x^{1/2})^2 = x$. For example, if $x = -4$, then $x^{1/2} = i2$, and $(x^{1/2})^2 = (i2)^2 = -4$, which is the original number. We have seen that two signals can be used to represent complex numbers electronically.

On the other hand, x^2 is not invertible[16] because both positive and negative values of x give positive values of x^2. However, *if we allow the existence of a second signal path* (as in the case of the square root), then x^2 becomes invertible. This is shown in the first half of the block diagram in Fig. 11.13. The second half of the diagram shows how this new kind of x^2 function, Inv x^2, can be inverted to recover the original value of x, including its sign.

In general, for any power a that is any real number, the invertible

[15] Also, sgn $x|x|^a$ is invertible.

[16] That is, $(x^2)^{1/2} \neq x$.

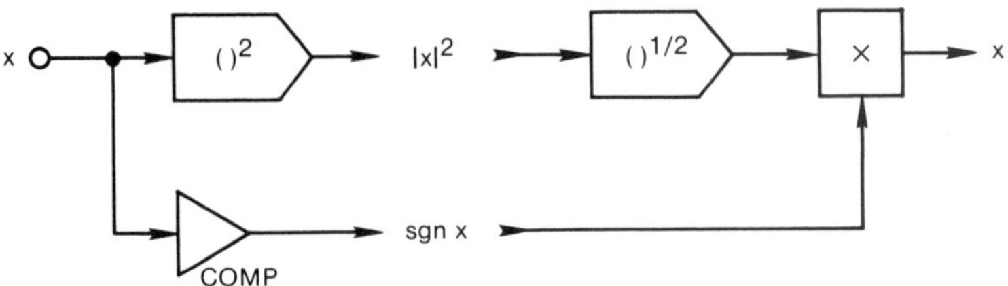

Figure 11.13 Block diagram of the invertible x^2 function and its inversion.

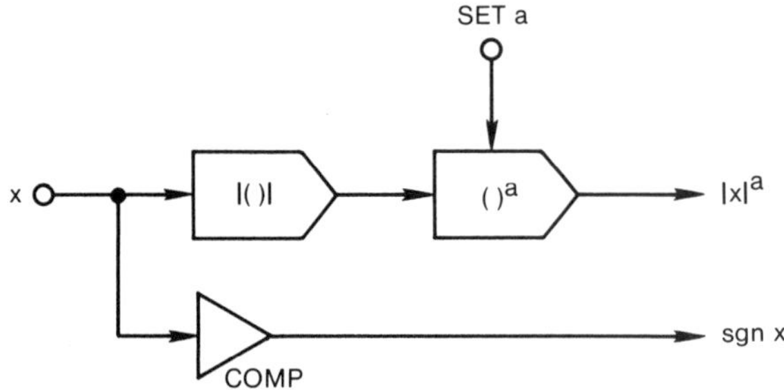

Figure 11.14 Block diagram of the invertible x^a function.

power is defined by the block diagram in Fig. 11.14. It is seen that the invertible power is a two-component vector that can be written

$$\text{Inv } x^a = \begin{bmatrix} |x|^a \\ \text{sgn } x \end{bmatrix} \tag{11.72}$$

With the invertible power, we have removed the "unfair advantage" that $( )^{1/2}$ had over $( )^2$ in its use of imaginary numbers for keeping track of sign. Indeed, all $( )^a$ for all real a-values have been put on an equal footing in that they are all single valued and invertible. The invertible power of x, while closely related to x^a, is potentially more useful than x^a because it carries more information. It consistently recognizes the vector nature of that function—something that x^a does not do when a is an integer.

An application of the invertible power is to analog signal encoding. The block diagram of such an encoder and its decoder is shown in Fig. 11.15. With this encoding system, a secure voice link can be established. By varying the parameter a, unauthorized decoding can be made extemely difficult. Or, by keeping a fixed at a prearranged value, the system can be simplified by the elimination of one transmission link. Refer to a similar discussion in subsection 7.1.2.

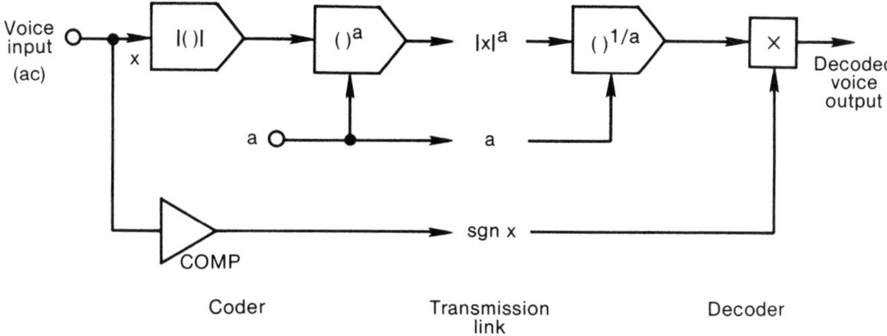

Figure 11.15 Block diagram of a secure voice communication system based on the invertible x^a function.

APPENDIX A: PROOFS OF THEOREMS AND LEMMAS

Definitions

In the material of this appendix, as throughout the text, derivatives are *complete derivatives* unless otherwise noted. For definitions of "complete derivative" and "intrinsic derivative," refer to Section 4.2.

The function $\text{saw}_a x$ is the sawtooth periodic function ("wave") defined by Eq. (3.20):

$$\text{saw}_a x = (a/\pi)\text{Tan}^{-1}\tan(\pi x/a)$$

The notation $-\frac{1}{2}a \underset{\leftarrow}{\le} x \underset{\rightarrow}{\le} \frac{1}{2}a$ means $-\frac{1}{2}a \leqslant x < \frac{1}{2}a$ for negative progression and $-\frac{1}{2}a < x \leqslant \frac{1}{2}a$ for positive progression.

The notations $f(-\frac{1}{2}a + 0)$ and $f(\frac{1}{2}a - 0)$ are the unilateral limits $\lim\limits_{x\to -a/2+} f(x)$ and $\lim\limits_{x\to a/2-} f(x)$, respectively.

First Fundamental Theorem (Sec. 3.5)

Given: **1.** $p(x)$ is periodic in x with period a.

2. $f(x) = p(x)$ in the interval $-\frac{1}{2}a \underset{\leftarrow}{<} x \underset{\rightarrow}{<} \frac{1}{2}a$.

$\to\to\to$ $p(x) = f(u)$

where $u = \mathit{saw}_a x$.

Proof: **1.** The interval $-\frac{1}{2}a \underset{\leftarrow}{\le} x \underset{\rightarrow}{\le} \frac{1}{2}a$ corresponds to one cycle of $p(x)$; $u = x$ in that interval. Therefore $f(u) = p(x)$ over at least one cycle, because of Given (2).

2. The function u is periodic with period a. Therefore $f(u)$ is periodic with period a.
3. Since $f(u)$ and $p(x)$ match over one cycle and their periods are the same, it follows that $f(u) = p(x)$ for all x-values.

Theorem 4.1 (Sec. 4.2)

When differentiating a function that contains no jumps, only the intrinsic portion of intermediate derivatives need be taken.

Outline of Proof

1. Delta pulses have no width; therefore any combination of a finite number of delta functions is still a delta function or zero.
2. Because of this, the delta-function components of intermediate derivatives cannot contribute to the intrinsic portion of the final derivative.
3. But the intrinsic derivative of a function with no jump is also its complete derivative.

Second Fundamental Theorem (Sec. 4.3)

Given: **1.** $p(x)$ is periodic in x with period a.

2. $f(x) = p(x)$ in the interval $-\frac{1}{2}a \underset{\leftarrow}{\le} x \underset{\rightarrow}{\le} \frac{1}{2}a$.

$$\rightarrow\rightarrow\rightarrow \quad \frac{d}{dx}p(x) = \frac{d}{du}f(u) + r\,\delta\left(\frac{a}{\pi}\cos\frac{\pi x}{a}\right)$$

Where: **1.** $u = \text{saw}_a x$

2. $r = f(-\frac{1}{2}a + 0) - f(\frac{1}{2}a - 0)$.

Proof: **1.** $u = x$ in the interval $-\frac{1}{2}a \underset{\leftarrow}{\le} x \underset{\rightarrow}{\le} \frac{1}{2}a$; therefore $\frac{d}{du}f(u) = \frac{d}{dx}p(x)$ in that interval.

2. u is periodic in x with period a; therefore $\frac{d}{du}f(u)$ is periodic in x with period a; and $\frac{d}{du}f(u)$ is the derivative of $p(x)$ for all real x, except for a possible delta pulse at $x = \frac{1}{2}ma$, where m is any odd integer.

3. A discontinuity exists in $p(x)$ at $x = \frac{1}{2}a$ of amount r; therefore the delta-pulse portion of the derivative of $p(x)$ at $x = \frac{1}{2}a$ is $r\,\delta(x - \frac{1}{2}a)$.

4. $\delta\left(\frac{a}{\pi}\cos\frac{\pi x}{a}\right)$ is a delta pulse train of period a with the "first" pulse ($m = 1$) occurring at $x = \frac{1}{2}a$; therefore the delta-pulse portion of the derivative of $p(x)$ at $x = \frac{1}{2}ma$ is $r\,\delta\left(\frac{a}{\pi}\cos\frac{\pi x}{a}\right)$, completing the proof.

Lemma 5.1 (Sec. 5.3)

If the average value of a periodic function with any period a is zero, then its integral function is also periodic with period a.

Proof: **1.** Given, $p(x)$ is periodic with period a and average value $\bar{p} = 0$.

2. $\bar{p} = \frac{1}{a}\int_x^{x+a} p(x)dx = \frac{1}{a} P(x)|_x^{x+a}$ by definition.

3. $P(x)|_x^{x+a} = P(x + a) - P(x) = 0$; therefore $P(x + a) = P(x)$. This is just the condition that $P(x)$ be periodic with period a.

Third Fundamental Theorem (Sec. 5.3)

Given: **1.** $p(x)$ is periodic in x with period a,

2. $f(x) = p(x)$ in the interval $-\frac{1}{2}a \underset{\leftarrow}{\le} x \underset{\rightarrow}{\le} \frac{1}{2}a$,

$\rightarrow\rightarrow\rightarrow$ $\int p(x)\, dx = \int f(u)\, du + \bar{p}\, \sigma_a(x)$

Where: **1.** $u = \mathrm{saw}_a x$

2. $\sigma_a(x) = x - u$

3. $\bar{p}$ is the average value of $p(x)$

Proof: **1.** $p(x) - \bar{p}$ is periodic, and $\int_{-a/2}^{a/2} [p(x) - \bar{p}]\, dx = 0$.

2. $\int[p(x) - \bar{p}]\, dx$ is periodic, by Lemma 5.1.

3. $\int[f(u) - \bar{p}]\, du = \int[p(x) - \bar{p}]\, dx$ over all x by the First Fundamental Theorem.

4. But $\int[f(u) - \bar{p}]\, du = \int f(u)\, du - \bar{p}\, u$ and $\int[p(x) - \bar{p}]\, dx = \int p(x)\, dx - \bar{p}\, x$.

5. From (3) and (4) we can obtain $\int p(x)\, dx = \int f(u)\, du + \bar{p}x - \bar{p}u$; Q.E.D.

Theorem 5.1 (Sec. 5.5)

Given: $\sum_{n=a}^{b} f(n) = f(a) + f(a + 1) + \ldots + f(b)$; a, b are integers, $a < b$.

$\rightarrow\rightarrow\rightarrow$ $\int_{a-1/2}^{b+1/2} f[\sigma_1(x)]\, dx = \sum_{n=a}^{b} f(n)$.

Proof: **1.** $\sigma_1(x)$ is a regular staircase function of unit rise and tread; therefore $\sigma_1(x) = n$ in the interval $n - \frac{1}{2} \underset{\leftarrow}{\le} x \underset{\rightarrow}{\le} n + \frac{1}{2}$.

2. $\int_{n-1/2}^{n+1/2} \sigma_1(x)\, dx = n$; therefore $\int_{n-1/2}^{n+1/2} f[\sigma_1(x)]\, dx = f(n)$.

3.
$$\sum_{n=a}^{b} f(n) = \sum_{n=a}^{b} \int_{n-1/2}^{n+1/2} f[\sigma_1(x)]\, dx$$
$$= \int_{a-1/2}^{b+1/2} f[\sigma_1(x)]\, dx;\ \text{Q.E.D.}$$

Lemma 9.1 (Sec. 9.1)

Given: $F(x)$ is continuous except for an infinite discontinuity at $x = a$; i.e., $\beta = -1$.

$\rightarrow\rightarrow\rightarrow$ $G(x) = \pm\ \mathrm{sgn}(x - a)\ F(x)$ has a break at infinity at $x = a$; i.e., $\beta = 1$

Where: $\beta = \lim_{x\to a} \dfrac{F(x)}{F(2a - x)}$.

Proof: **1.** $\lim_{x\to a} \dfrac{F(x)}{F(2a - x)} = -1$, given.

2. $$\lim_{x\to a} \frac{G(x)}{G(2a - x)}$$
$$= \lim_{x\to a} \frac{\pm\ \mathrm{sgn}(x - a)F(x)}{\pm\ \mathrm{sgn}(a - x)F(2a - x)}$$
$$= \lim_{x\to a} \frac{-F(x)}{F(2a - x)} = -\lim_{x\to a} \frac{F(x)}{F(2a - x)} = 1, \text{ Q.E.D.}$$

Theorem 9.1 (Sec. 9.2)

Given: $F(x)$ is continuous except for an infinite discontinuity at $x = a$; i.e., $\beta = -1$. Multiple values of a are allowed.

$\rightarrow\rightarrow\rightarrow$ $$\frac{d}{dx} F(x) = \phi(x) + 2\ \mathrm{sgn}\ uF(x)\ \delta(u) \frac{du}{dx}$$

Where: **1.** $\phi(x)$ is the intrinsic derivative of $F(x)$

2. $u = u(x)$ is any continuous function with $u(a) = 0$, $u(x) \neq 0$ when $x \neq a$, and with unit zero-crossing slope.

3. $\beta = \lim_{x\to a} \dfrac{F(x)}{F(2a - x)}$.

Proof: **1.** $\lim_{x\to a} \dfrac{F(x)}{F(2a - x)} = -1$ from Eq. (9.1).

2. Define $G(x) = \mathrm{sgn}\ u\ F(x)$. The function $u = u(x)$ is such that $\lim_{x\to a} \dfrac{G(x)}{G(2a - x)} = 1$; i.e., the infinite discontinuity in $F(x)$ becomes a break at infinity in $G(x)$ by Lemma 9.1.

3. $$\frac{d}{dx} F(x) = \frac{d}{dx} [\mathrm{sgn}\ u\ G(x)] = \mathrm{sgn}\ u \frac{d}{dx} G(x) + G(x) \frac{d}{dx} \mathrm{sgn}\ u.$$

4. But
$$\frac{d}{dx} G(x) = \mathrm{sgn}\ u\ \phi(x) \text{ by Theorem 4.1, giving}$$
$$\frac{d}{dx} F(x) = \phi(x) + \mathrm{sgn}\ u\ F(x) \frac{d}{dx} \mathrm{sgn}\ u; \text{ Q.E.D.}$$

APPENDIX B: POWERS AND DERIVATIVES OF THE DIRAC DELTA FUNCTION

The Amplitude of δ and Its Reciprocal

The amplitude of $\delta(x)$ is defined as its "value" at $x = 0$. This is written symbolically as $\delta(0)$. We note that $\delta(0)$ is an *infinite object,* analogous to a constant. It is a *particular infinity*.

Widder[1] pointed out that the reciprocal of an infinity is an infinitesimal. We now proceed to show that the reciprocal of $\delta(0)$, written $[\delta(0)]^{-1}$, is the infinitesimal $\delta^{-1}(0)$, where

$$\delta^{-1}(x) = x^2\delta(x) \tag{B1}$$

from Eq. (9.2). By "reciprocal" we obviously mean

$$\delta(0)[\delta(0)]^{-1} = \left[\frac{dS}{dx}\right]_{x=0}\left[\frac{dx}{dS}\right]_{x=0} = 1 \tag{B2}$$

Proof that $[\delta(0)]^{-1} = \delta^{-1}(0)$ follows.

In Section 4.2 it is stated that the classical definition of derivative as the limit of $\Delta y/\Delta x$ does not readily lend itself to finding the derivative at a discontinuity. Nevertheless, there must be no contradiction between that definition of derivative and whatever other definition we choose in a particular situation. This being so, we are obliged to equate $dS(x)/dx$

[1] D. V. Widder, *Advanced Calculus* (Englewood Cliffs, N.J.: Prentice-Hall, Inc., 1947), p. 231.

and the limit of $\Delta S(x)/\Delta x$, or their reciprocals, and accept whatever this leads to.

We take Δx at $x = 0$ so that it straddles $x = 0$, and recognizing that $\Delta S(x)$ with Δx so defined is $+1$ when Δx is positive (positive progression) and -1 when Δx is negative (negative progression), we have

$$[\delta(0)]^{-1} = \left[\frac{dx}{dS(x)}\right]_{x=0} = \left[\lim_{\Delta x \to 0} \frac{\Delta x}{\Delta S(x)}\right]_{x=0} \tag{B3a}$$

$$[\delta(0)]^{-1} = \lim_{\Delta x \to 0} |\Delta x| \tag{B3b}$$

The situation is depicted in Fig. B1.

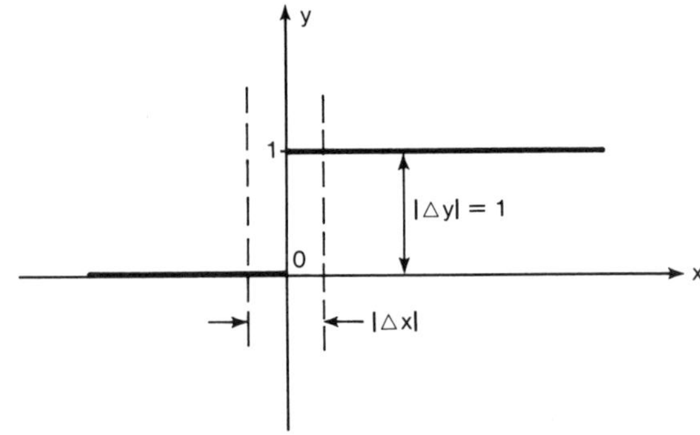

Figure B.1 Illustration of Δx and Δy at $x = 0$ for the Heaviside step function.

The index of a limit can be changed if the quantity is changed accordingly. Therefore we can write Eq. (B3b) as

$$[\delta(0)]^{-1} = \lim_{x \to 0} |x| \tag{B3c}$$

From Eqs. (B2) and (B3c) we have

$$\lim_{x \to 0} |x|\,\delta(0) = [|x|\,\delta(x)]_{x=0} = 1 \tag{B4a}$$

But

$$|x|\,\delta(x) = \delta^0(x) \tag{B4b}$$

from Eq. (9.2), so that Eq. (B4a) becomes

$$\rightarrow\rightarrow\rightarrow \quad \delta^0(0) = 1 \tag{B5}$$

Since $\delta^0(x)$ is zero when $x \neq 0$, then $\delta^0(x)$ can take on only the two values 0 and 1, and it is apparent that for $n > 0$

$$\rightarrow\rightarrow\rightarrow \quad [\delta^0(x)]^n = \delta^0(x) \tag{B6}$$

Consequently, we can write

$$|x|\,\delta(x) = [|x|\,\delta(x)]^2 = [x^2\,\delta(x)]\,\delta(x) = \delta^{-1}(x)\,\delta(x) \tag{B7}$$

so that

$$\delta^{-1}(0)\,\delta(0) = 1 \tag{B8a}$$

Finally, from Eqs. (B2) and (B8a) we have

$$\rightarrow\rightarrow\rightarrow \quad [\delta(0)]^{-1} = \delta^{-1}(0) \tag{B8b}$$

We now define the amplitude of a general delta function. Let $f(x)$ be single valued and unipolar at $x = a$. The amplitude of the delta function $f(x)\,\delta(u)$ is defined as $f(a)\,\delta(0)$ where a is the value of x for which $u = 0$. Further, $f(a)$ is called the *amplitude factor*. Examples follow.

Example B1. The amplitude factor of a particular delta function is given as $\delta^{-1}(0)$. It follows that the amplitude of that delta function is 1.

Example B2. Let $\Delta(x)$ be the "bipolar" delta function $x\delta(x)$, so that $\Delta(0 + 0) = 1$ and $\Delta(0 - 0) = -1$. The amplitude of this delta function is undefined because $f(x) = x$ is not unipolar at $x = 0$.

Example B3. Since $\delta^0(0) = 1$, it follows that

$$\delta^0(0)\,\delta(x) = \delta(x)$$

Extent of a Delta Pulse

It is pointed out in Chapter 2 that the relationship, with $u = u(x)$,

$$\delta(u) = |du/dx|^{-1}\,\delta(x - a) \tag{(B9)}$$

is true only in the sense that the contents of the two equated delta pulses are equal. It is recalled that a is such that $u = 0$ when $x = a$. We now point out what the differences are in the delta pulses themselves.

First, the contents of both delta pulses (relative to x) are the same, namely, $[|du/dx|^{-1}]_{x=a}$.

Second, the amplitude of the left-hand delta pulse of Eq. ((B9)) is $\delta(0)$, whereas the amplitude of the right-hand one is $\delta(0)[|du/dx|^{-1}]_{x=a}$. So the two pulse amplitudes are different by the factor $[|du/dx|^{-1}]_{x=a}$.

Finally we define the *extent* of a delta pulse as its content times the reciprocal of its amplitude. For example, the extent of $\delta(x)$ can easily be shown to be $\delta^{-1}(0)$. The extent of the left-hand delta pulse of Eq. ((B9)) is $\delta^{-1}(0)[|du/dx|^{-1}]_{x=a}$, whereas the extent of the right-hand one is $\delta^{-1}(0)$. That is, the two pulse extents are different by the factor $|du/dx|_{x=a}$.

In summary, the contents of the two delta pulses of Eq. ((B9)) are the same but their amplitudes and extents are different. To say that the two delta pulses are "equal" is a little like saying that the two rectified cosinewaves $|\cos x|$ and $2|\cos \frac{1}{2}x|$ are "equal" because their integrals over one cycle are the same.

As an example of the above relationships, we let $u = 2x$ so that $|du/dx| = 2$. We see that the amplitude and extent of $\delta(u)$ are $\delta(0)$ and $\frac{1}{2}\,\delta^{-1}(0)$, respectively, and the amplitude and extent of $|du/dx|^{-1}\,\delta(x - a)$ are $\frac{1}{2}\delta\,(0)$ and $\delta^{-1}(0)$ respectively. In this case $a = 0$.

In this example, the content of both delta pulses is $\frac{1}{2}$, as required.

In summary, the delta function on the right side of Eq. ((B9)) *does not* equal the delta function on the left side of that equation, even though the contents of the two delta pulses are the same.

In general, we have the following . . .

Rule B1: For two delta functions to be *equal,* it is necessary and sufficient that the quantity, positions, amplitudes and contents of their delta pulses be equal. Examples follow.

Example B4. In this example we show that the extent of $A\delta^0(x)$ is $\delta^{-1}(0)$. The content of $A\delta^0(x)$ is $A\delta^{-1}(0)$ and its amplitude is $A\delta^0(0) = A$. Therefore its extent is content/amplitude $= A\delta^{-1}(0)A^{-1} = \delta^{-1}(0)$.

Example B5. It can be shown that $\delta^0(x)\,\delta(0) = \delta(x)$ by Rule B1.

Example B6. By Rule B1 we have

$$[\delta(x - a)]^2 = \delta(0)\,\delta(x - a)$$

Therefore, when $b = a$, $\delta(x - a)\,\delta(x - b) = \delta(0)\,\delta(x - a)$ and when $b \neq a$, $\delta(x - a)\,\delta(x - b) = 0$. It follows that, in general,

$$\delta(x - a)\,\delta(x - b) = \delta(a - b)\,\delta(x - a)$$

Therefore

$$\int \delta(x - a)\,\delta(x - b)\,dx = \delta(a - b)S(x - a) + C$$

Positive Powers of δ

We now proceed to show that for $n > 0$, the nth power of $\delta(x)$ is

$$[\delta(x)]^n = \delta^n(x) \qquad \text{(B10)}$$

where $\delta^n(x)$ is given by Eq. (9.2). We write

$$\delta(x) = |x|^{-1}|x|\,\delta(x) \qquad \text{(B11)}$$

where $|x|\;\delta\,(x)$ is a delta pulse of unit amplitude. Raising both sides of Eq. (B.11) to the nth power, we obtain, for $n > 0$,

$$[\delta(x)]^n = |x|^{-n}|x|\,\delta(x) \qquad \text{(B12)}$$

because the nth power of $\delta^0(x)$ is just $\delta^0(x)$.

Equation (B12) becomes

$$\rightarrow\rightarrow\rightarrow \quad [\delta(x)]^n = |x|^{-n+1}\,\delta(x) = \delta^n(x) \qquad \text{(B13)}$$

In the above proof, the only stipulation we have made concerning n is that it must be positive. It can be any positive real number.

We note that in all the equations of this section, there is a delta function on both sides. In general, when writing equations involving delta functions it is necessary that this be so.

This section deals only with positive powers of $\delta(x)$. Negative and zero powers have no meaning at this point. However, negative and zero *orders* of $\delta(x)$ do have meaning. They are defined by Eq. (9.2).

As a final point, we note that some physical applications require the use of the *three-dimensional delta function,* $\delta(x, y, z)$ defined as $\delta(x)\,\delta(y)\,\delta(z)$.[2] Using this notation we can also write $\delta^2(x)$ as $\delta(x, x)$, $\delta^3(x)$ as $\delta(x, x, x)$, and so on. Examples follow.

Example B7. We show that $\delta^0(x)\,\delta(x) = \delta(x)$. Write

$$\delta^0(x)\,\delta(x) = |x|[\delta x)]^2 = |x|\,\delta^2(x) = |x||x|^{-1}\,\delta(x) = \delta(x)$$

Example B8. We show that $\delta^1(x) = \delta(x)$. Write[3]

$$\delta^1(x) = |x|^0\,\delta(x) = \delta(x)$$

Example B9. It is easy to show that the nth order of $\delta(x)$ can also be written

$$\delta^n(x) = |x|^{-n}\,\delta^0(x)$$

Example B10. It is also easy to show that

$$\delta^a(x)\,\delta^b(x) = \delta^{a+b}(x)$$

for all real values of a and b.

Example B11. From Example B6 and the relationship $[\delta(\cos x)]^2 = \delta^2(\cos x)$, we can "evaluate" the integral of Eq. (9.6) in Chapter 9 to give

$$\int_0^{\pi} \delta^2(\cos x)\,dx = \delta(0)$$

The First Derivative of δ

We write from Eq. (9.2) with $n = 0$,

$$\delta^0(x) = |x|\,\delta(x) \tag{B14}$$

and differentiate to get

$$|x|\frac{d}{dx}\delta(x) = \frac{d}{dx}\delta^0(x) - \operatorname{sgn} x\,\delta(x) \tag{B15}$$

[2] $\delta(x, y, z)$ may be written $\delta(\mathbf{r})$ where $\mathbf{r}$ denotes the vector (x, y, z). Some authors write $\delta^3(\mathbf{r})$ for the three-dimensional delta function.

[3] $|x|^\circ$ is an indeterminate expression of the form 0° when $x = 0$. Nevertheless, it can be shown by the method of limits to equal 1 for all x-values.

We can also differentiate

$$\delta^2(x) = |x|^{-1}\,\delta(x) \tag{B16}$$

to get

$$|x|\frac{d}{dx}\,\delta(x)[2\,\delta(x) - |x|^{-1}] = -\delta(x)/x \tag{B17}$$

Substituting for $|x|\dfrac{d}{dx}\,\delta(x)$ from Eq. (B15) into Eq. (B17) and simplifying, we obtain

$$\frac{d}{dx}\,\delta^0(x)[2\,\delta^0(x) - 1] = 0 \tag{B18}$$

Of the two factors in Eq. (B18), $2\,\delta^0(x) - 1$ is either $+1$ (when $x = 0$) or -1 (when $x \neq 0$). This means that

$$\rightarrow\rightarrow\rightarrow \quad \frac{d}{dx}\,\delta^0(x) = 0 \tag{B19}$$

We can subsequently obtain without difficulty the expression

$$\rightarrow\rightarrow\rightarrow \quad \frac{d}{dx}\,\delta(x) = -\delta(x)/x \tag{B20}$$

The distribution $-\delta(x)/x$ is a *bipolar delta function.* (A bipolar delta pulse is called a *doublet.*) This is intuitively "right" because the derivative of a wide pulse is a bipolar pulse. Examples follow.

Example B12. The second derivative of the sine-derived squarewave is

$$\frac{d^2}{dx^2}\,\text{sqr}\,x = -2\,\delta(\sin x)/(\sin x)$$

Example B13. The derivative of $\text{ln}|x|\;\delta(x)$ is

$$\frac{d}{dx}\,[\text{l}n|x|\;\delta(x)] = (\text{L}n|x| - 1)\frac{d}{dx}\,\delta(x)$$

Example B14. The integral of $\delta(x)$ sgn x is

$$\int \delta(x)\,\text{sgn}\,x\,dx = \text{L}n|x|[\delta^0(x) + C] + C'$$

Example B15. The integral of $\delta^n(x)$ sgn x is

$$\int \delta^n(x)\,\text{sgn}\,x\,dx = \delta^0(x)\int |x|^{-n}\text{sgn}\,x\,dx + C$$

Example B16. The definite integral over all x of $\delta(x)S(x)$ is

$$\int_{-\infty}^{\infty} \delta(x)S(x)\,dx = \tfrac{1}{2}$$

Higher Derivatives of δ

We write for the second derivative of $\delta(x)$

$$\frac{d^2}{dx^2}\,\delta(x) = \frac{d^2}{dx^2}\,[|x|^{-1}\,\delta^0(x)] \tag{B21a}$$

$$= \delta^0(x)\,\frac{d^2}{dx^2}\,|x|^{-1} \tag{B21b}$$

$$= \delta^0(x)\,\frac{d}{dx}\,[-x^{-2}\mathrm{sgn}\,x] \tag{B21c}$$

$$= \delta^0(x)[2x^{-3}\mathrm{sgn}\,x - 2x^{-2}\,\delta(x)] \tag{B21d}$$

Finally,

$$\frac{d^2}{dx^2}\,\delta(x) = 2\,\delta(x)x^{-2} - 2\,\delta(x)x^{-2} = 0 \tag{B21e}$$

Thus, *the second derivative of* $\delta(x)$ *is zero*! As a result, all higher derivatives are also zero. But the story does not end here.

Proceeding from Eq. (B21b), we define the nth *formal derivative of* $\delta(x)$, written $\delta^{(n)}(x)$, as

$$\rightarrow\rightarrow\rightarrow \quad \delta^{(n)}(x) = \delta^0(x)\left[\frac{d^n}{dx^n}\,|x|^{-1}\right]_{\mathrm{IN}} \tag{B22}$$

where $[\ ]_{\mathrm{IN}}$ indicates the intrinsic part of the derivative inside the brackets.

Letting $n = 1$, we see that the first formal derivative of $\delta(x)$, written $\delta'(x)$, is just its derivative as found earlier:

$$\delta'(x) = \frac{d}{dx}\,\delta(x) = -\delta(x)/x \tag{B23}$$

It is also easy to show that the second formal derivative of $\delta(x)$, written $\delta''(x)$, is

$$\delta''(x) = 2\,\delta(x)x^{-2} \tag{B24}$$

In general, it is not difficult to show that the nth formal derivative of $\delta(x)$ is

$$\rightarrow\rightarrow\rightarrow \quad \delta^{(n)}(x) = (-1)^n n!\,\delta(x)x^{-n},\ n = 0, 1, 2, \ldots \tag{B25}$$

The case $n = 0$ is included in Eq. (B25) because it gives the desired expression for the zeroth derivative: i.e., $\delta^{(0)}(x) = \delta(x)$.

Example B17. For $n = 0, 2, 4, 6, \ldots$

$$\delta^{(n)}(x) = n!\,\delta^{n+1}(x)$$

Example B18. For $n = 1, 3, 5, 7, \ldots$

$$\delta^{(n)}(x) = n!\,\delta^{n-1}(x)\,\delta'(x)$$

Example B19. Let $(d/dx)F(x) = f(x)$ exist and be single valued at $x = 0$. Further, let $F(x) = 0$. We write

$$\int F(x)\, \delta'(x)\, dx = -\int F(x)\, \delta(x)/x\, dx$$
$$= -F(x)/x|_{x=0} S(x) + C$$

Further, we write

$$f(x) = \lim_{\Delta x \to 0} \frac{F(x + \Delta x) - F(x)}{\Delta x}$$

and

$$f(0) = \lim_{\Delta x \to 0} \frac{F(\Delta x)}{\Delta x} = \left. \frac{F(x)}{x} \right|_{x=0}$$

Finally, we have

$$\int F(x)\, \delta'(x)\, dx = -f(0)S(x) + C$$

It is customary to define formal derivatives of δ in terms of such integrals.[4]

A "Universal" Delta-Sequence Function

A function that approaches δ in the limit is called a *delta-sequence function*. One such function is suggested by the relationship $\delta(x) = |x|^{-1}\, \delta^0(x)$.

Consider the wide nonperiodic pulse depicted in Fig. B2. The pulse amplitude is $1/b$ and the pulse width is $2a$.

We want the extent to be b and the content to be 1; therefore the relationship between a and b must be $b = a(e^{1/2} - 1)^{-1}$, as can be shown from the relationship

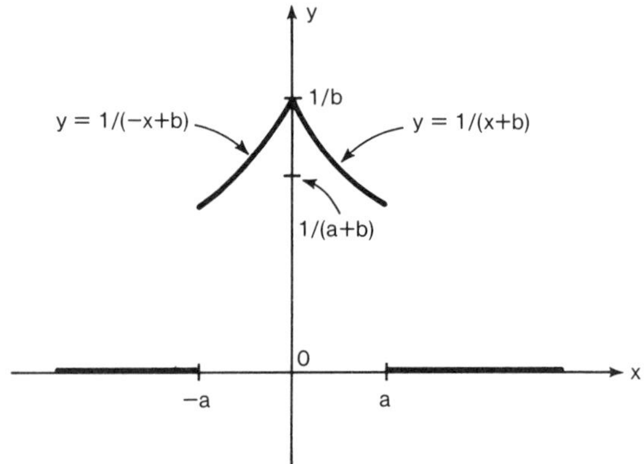

Figure B.2 A "universal" delta-sequence function, $y = \delta_a(x)$.

[4]See for example G. A. Korn and T. M. Korn, *Mathematical Handbook for Scientists and Engineers* (New York: McGraw-Hill, 1961), p. 743.

$$\tfrac{1}{2} = \int_0^a (x + b)^{-1}\, dx = \ln(a + b) - \ln b \tag{B26}$$

where $e = 2.71828\ldots.$

We write $\delta_{0+}(x) \approx \delta(x)$, where $\delta_a(x) \to \delta_{0+}(x)$ as $a \to 0+$, and

$$\to\to\to \quad \delta_a(x) = \frac{1}{b + |x|}\,\mathrm{puls}^{a}_{-a}x. \tag{B27}$$

The function $\delta_{0+}(x)$ defines a pulse having infinite amplitude, infinitesimal width, unit content, an indefinite integral of $S(x) + C$, and having correct functions for powers and derivatives.

It is because of these several properties that we refer to $\delta_a(x)$ as defined in Eq. (B27) as a "universal" delta-sequence function.

Referring to Fig. B2, we note that the "eaves" of the pulse (i.e., the low points of the curves) are at a fixed fraction of the distance to the peak (namely, $\delta_a(a)/\delta_a(0) = e^{-1/2} = 0.61$) regardless of the value of a. Therefore it can legitimately be considered that Fig. B2 shows a "true picture" of δ that has been subjected to an infinite linear magnification in the horizontal direction and an infinite linear minification in the vertical direction.

APPENDIX C: PROPERTIES OF THE NULL FUNCTION δ^0

From Eq. (9.2), the delta function $\delta^0(x)$ is defined as $|x|\ \delta(x)$. Several names characterize δ^0 because it has several useful and unique properties.

1. $\delta^0(x)$ is a *null function* because its definite integral over any interval is zero.

2. δ^0 is a *mock constant* because its derivative is identically zero.

3. δ^0 is the *identity delta function* because $\delta^0(x)\Delta = \Delta$ where Δ is any delta function with pulse at $x = 0$. An example is $\delta^0(x)\delta(x) = \delta(x)$.

4. δ^0 is a *delta-function generator* because $\delta^0(x)f(x) = \Delta$ where f is any function and Δ is the resulting delta function. An example is $\delta^0(x)|x|^{-1} = \delta(x)$.

5. δ^0 is a *binary inverter* because it changes ones to zeros and zeros to ones. That is, $\delta^0(x) = 1 - S$ where S is zero or one.

6. δ^0 is an *unblanking-function generator* because it changes nonzeros to zero and zero to one. An example is given in Eqs. (11.32), namely, $z = \delta^0[\text{Im}\ f(x)]$.

7. δ^0 is a *coincidence detector* because when $a = b$, $\delta^0(a - b) = 1$; otherwise, it is zero.

APPENDIX D: PRINCIPAL VALUE OF AN ANGLE

Consider the function

$$\Theta = \text{saw}_{2\pi}\theta \tag{D1}$$

This function varies between $-\pi$ and $+\pi$ in a linear sawtooth manner as θ varies between $-\infty$ and $+\infty$. If θ is interpreted as an angle, we say that Θ is the *principal value* of that angle.

From Eq. (3.20), Eq. (D1) can be written

$$\begin{aligned} \Theta &= 2\,\text{Tan}^{-1}\tan(\tfrac{1}{2}\theta) \\ &= 2\,\text{Tan}^{-1}[\sin\theta/(1+\cos\theta)] \end{aligned} \tag{D2}$$

Finally, if we let $\cos\theta = x/r$ and $\sin\theta = x'/r$, Eq. (D2) becomes

$$\rightarrow\rightarrow\rightarrow \quad \Theta = 2\,\text{Tan}^{-1}[x'/(r+x)] \tag{D3}$$

Equation (D1) can also be written

$$\begin{aligned} \Theta &= 2\,\text{Tan}^{-1}[(1-\cos\theta)/\sin\theta] \\ \rightarrow\rightarrow\rightarrow \quad &= 2\,\text{Tan}^{-1}[(r-x)/x'] \end{aligned} \tag{D4}$$

From Eq. (3.37) we have

$$\text{saw}_{2\pi}\theta - \tfrac{1}{2}\pi\,\text{sqr}\,\theta = \text{saw}_{\pi}\theta - \tfrac{1}{2}\pi\,\text{sqr}(2\theta). \tag{D5}$$

Therefore Θ can also be written

$$\rightarrow\rightarrow\rightarrow \quad \Theta = \text{Tan}^{-1}(x'/x) + \tfrac{1}{2}\pi\,\text{sgn}\,x' - \tfrac{1}{2}\pi\,\text{sgn}(x\,x') \tag{D6}$$

We note that $\sin \Theta = \sin \theta$ and $\cos \Theta = \cos \theta$, from Corollary 1 to the First Fundamental Theorem.

In summary, when considering the polar coordinate system r, Θ, the polar angle Θ is single valued even though θ is multiple valued. For example, if $\Theta = 0$, then $\theta = \ldots, -2\pi, 0, 2\pi, 4\pi, 6\pi, \ldots$. The polar coordinates r, Θ refer to a Euclidean plane, whereas the polar coordinates r, θ refer to an infinitely layered *Riemann surface*.

APPENDIX E: THE PARALLEL OPERATION

In the electronics literature, one occasionally encounters the *parallel operation,* $\|$, in addition to the common arithmetic operations of addition, subtraction, multiplication, and division. It means

$$a \| b = \frac{ab}{a + b} \tag{E1}$$

It is apparent that the parallel operation is commutative so that

$$b \| a = a \| b \tag{E2}$$

and associative so that

$$\begin{aligned}(a \| b) \| c &= a \| (b \| c) \\ &= \frac{abc}{ab + ac + bc} = a \| b \| c \\ &= b \| c \| a = c \| a \| b \end{aligned} \tag{E3}$$

etc.
Other properties are

$$a \| a = a/2 \tag{E4}$$

$$a \| a \| a = a/3 \tag{E5}$$

$$a \| a \| \ldots \| a_{(n \text{ times})} = a/n \tag{E6}$$

The result of applying the addition operation to a set of numbers is called the *total value*. Similarly, we will call the result of applying the parallel operation to a set of numbers the *focal value*.

The resistance of a network composed of resistors as seen external to the network is called the *equivalent resistance* of the network. For a series string, this is the total value of the resistances. For a parallel string it is the focal value of resistances. This and other examples follow. The equivalent resistance of two resistors in parallel is the focal value R_f, with

$$R_f = R_1 \parallel R_2 \tag{E7}$$

where R_1 and R_2 are the resistances of the individual resistors. The equivalent capacitance of two capacitors *in series* is

$$C_f = C_1 \parallel C_2 \tag{E8}$$

where C_1 and C_2 are the individual capacitances. The focal value (focal length) f of a thin lens is

$$f = p \parallel q \tag{E9}$$

where p and q are the image and object distances. The Q of a parallel or series LC resonant circuit is

$$Q = Q_L \parallel Q_c \tag{E10}$$

where Q_L is the Q of the inductor (L) and Q_c is the Q of the capacitor (C) at the resonant frequency. Finally, the equivalent conductance of two resistors *in series* is

$$G_f = G_1 \parallel G_2 \tag{E11}$$

where G_1 and G_2 are the conductances of the individual resistors.

The parallel operation takes its name from the application of Eq. (E7). The term "focal value" is taken from the relationship of Eq. (E9).

Just as subtraction is the inverse of the addition operation (so that if $T = a + b$ then $a = T - b$), and division is the inverse of the multiplication operation (so that if $T = ab$ then $a = T/b$), there is also an inverse of the parallel operation.

If $T = a \parallel b$, then we might write $a = T \mathbin{//} b$, but actually we find that

$$a = T \parallel (-b) \tag{E12}$$

so that a special symbol is not needed as is the case with subtraction and division. It is also true that

$$a = -[(-T) \parallel b] \tag{E13}$$

Further, if $T = a \parallel b \parallel c$, then

$$a = T \parallel (-b) \parallel (-c) \tag{E14}$$

$$b = T \parallel (-c) \parallel (-a) \tag{E15}$$

$$c = T \parallel (-a) \parallel (-b) \tag{E16}$$

Alternatively we can write

$$-a = (-T) \parallel b \parallel c \tag{E17}$$

$$-b = (-T) \parallel c \parallel a \tag{E18}$$

$$-c = (-T) \parallel a \parallel b \tag{E19}$$

Other relationships that can be derived from the inverse operation are:

$$(-a) \parallel b = -[a \parallel (-b)] \tag{E20}$$

$$a \parallel b \parallel (-b) = a \tag{E21}$$

$$a \parallel (\pm\infty) = a \tag{E22}$$

$$a \parallel (-a) = -[a \parallel (-a)] = \pm\infty \tag{E23}$$

$$(-a) \parallel (-b) = -(a \parallel b) \tag{E24}$$

$$a = (a/2) \parallel (-a) \tag{E25}$$

$$a = (a/3) \parallel (-a) \parallel (-a) \tag{E26}$$

Analog instrumentation of the parallel operation is represented by the block diagram of Fig. E1.

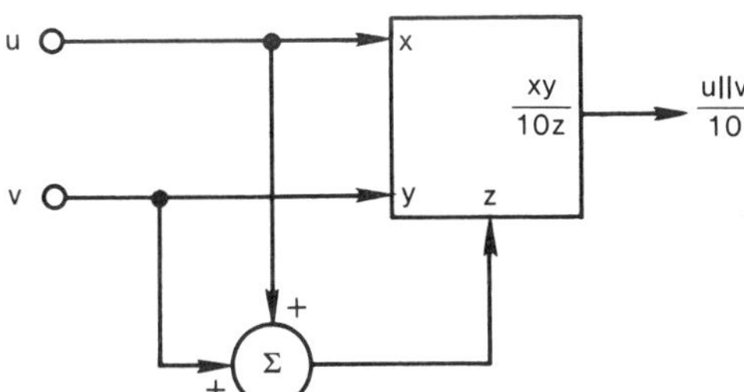

Figure E.1 Block diagram of the parallel operation.

If you program your computer to do the parallel operation, then you can easily find the equivalent resistance of resistors in parallel. You can also use that same program to find the resistance required to parallel a given resistance to obtain a desired equivalent resistance by use of Eq. (E11) or (E12).

The parallel operation can be programmed into the Hewlett-Packard model 67 and 97 scientific calculators by the following simple program:

LBLA/STO1/X ⇄ Y/STO + 1/ × /RCL1/ ÷ /RTN

Examples of the use of the inverse parallel operation follow.

Question. What value of resistor must be paralleled with a 150-kΩ resistor so that the combination has a resistance of 53 kΩ?

Answer. $R_f = 53$, $R_1 = 150$, $R_2 = R_f \parallel (-R_1) = 53 \parallel (-150) = 82$, so that an 82-kΩ resistor paralleled with a 150-kΩ resistor will produce 53 kΩ.

Question. What is the image distance for the proper focus of an object that is 150 cm from a thin lens whose focal length is 25 cm?

Answer. $f = 25$, $p = 150$, $q = f \parallel (-p) = 25 \parallel (-150) = 30$. So the correct answer is 30 cm.

Figure E2 is a nomograph that performs the parallel operation. Two self-explanatory examples of its use are given there.

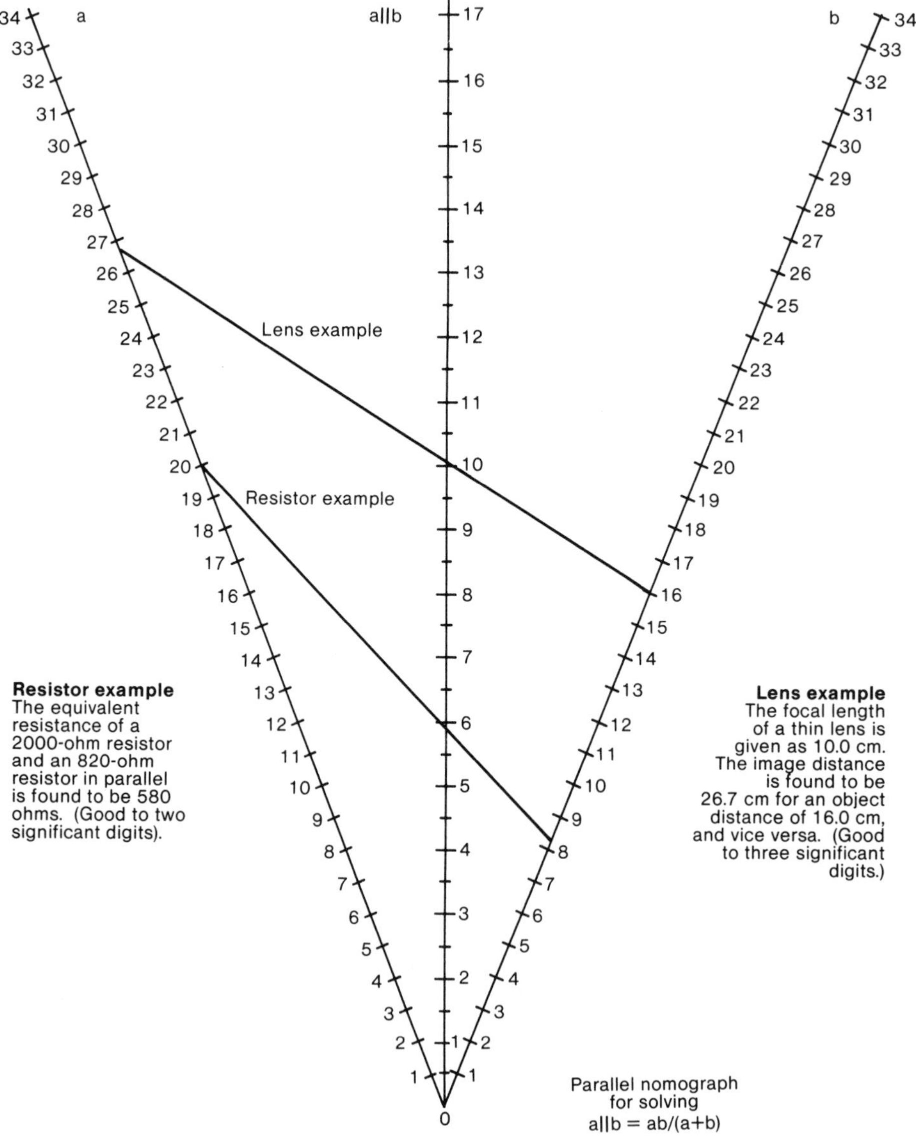

Figure E.2 Parallel nomograph for solving $a \parallel b = ab/(a + b)$.

APPENDIX F: CURVES

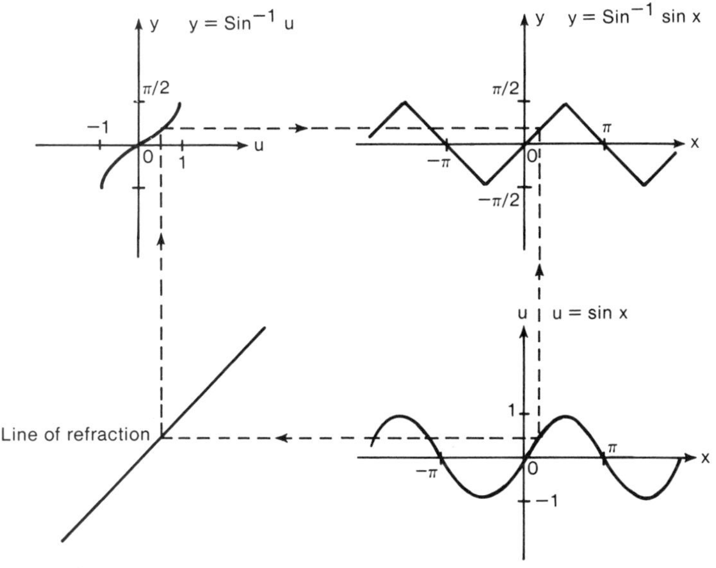

Figure F.1 $y = \mathrm{Sin}^{-1} \sin x$.

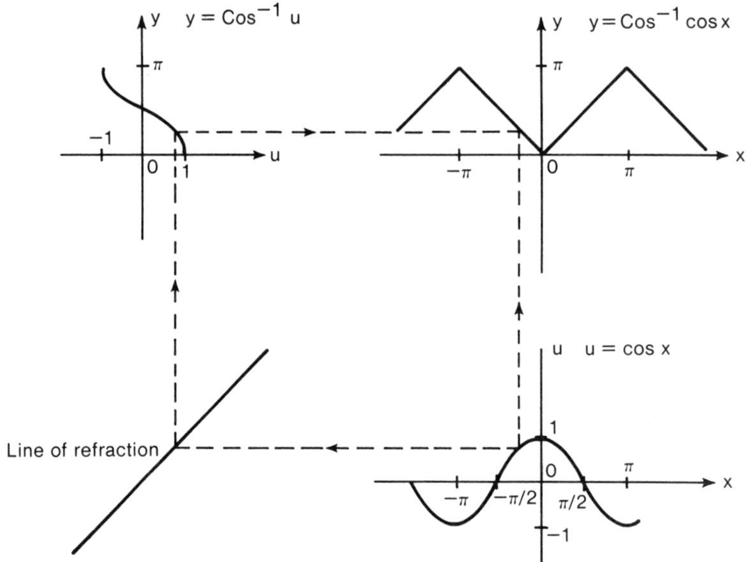

Figure F.2 $y = \mathrm{Cos}^{-1} \cos x$.

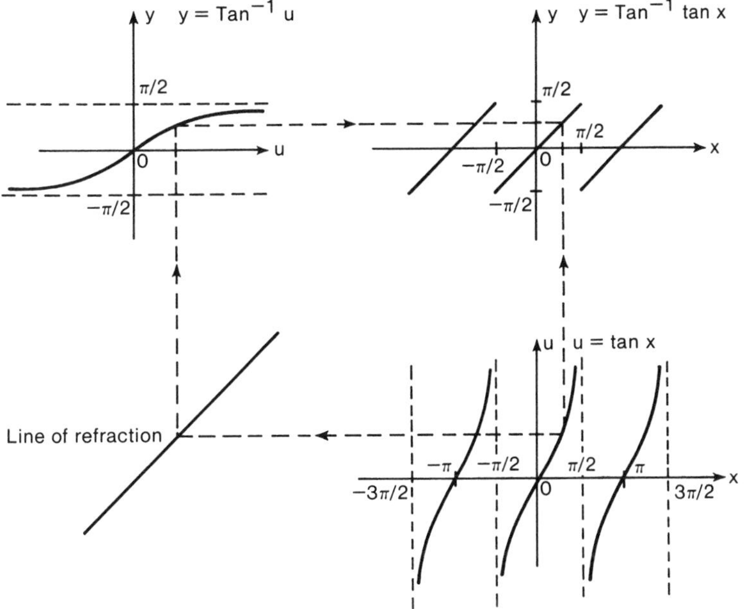

Figure F.3 $y = \text{Tan}^{-1} \tan x$.

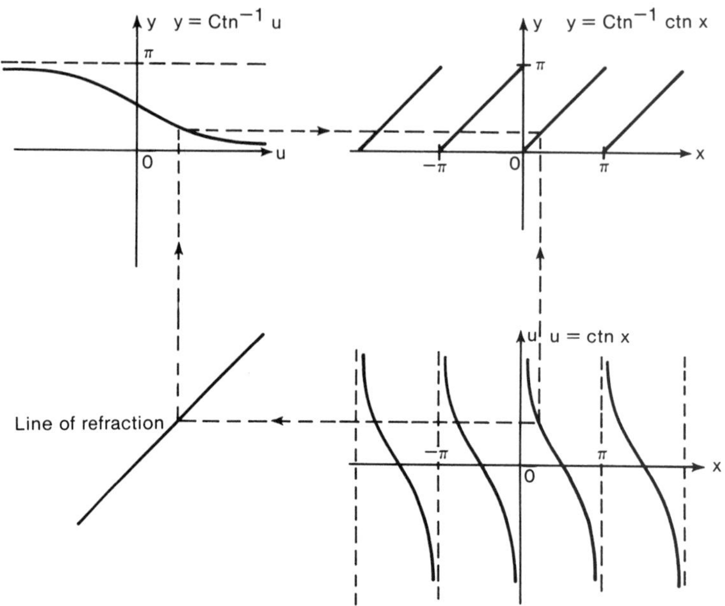

Figure F.4 $y = \text{Ctn}^{-1} \text{ ctn } x$.

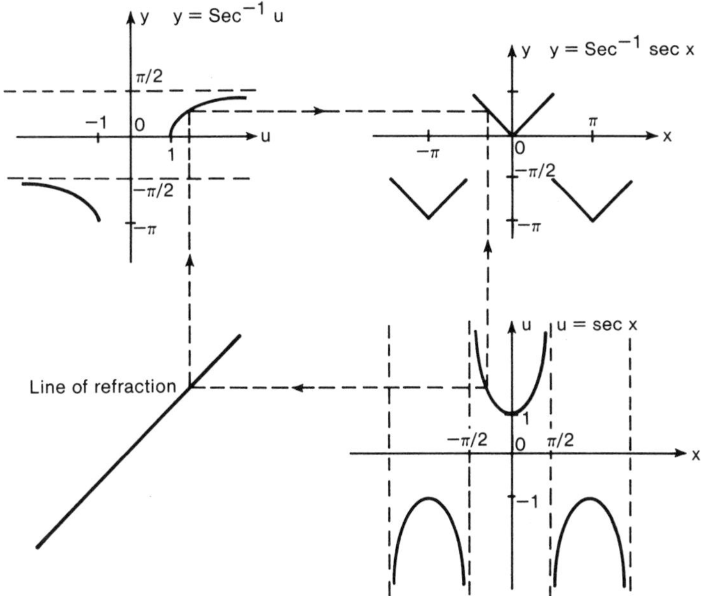

Figure F.5 $y = \text{Sec}^{-1} \sec x$.

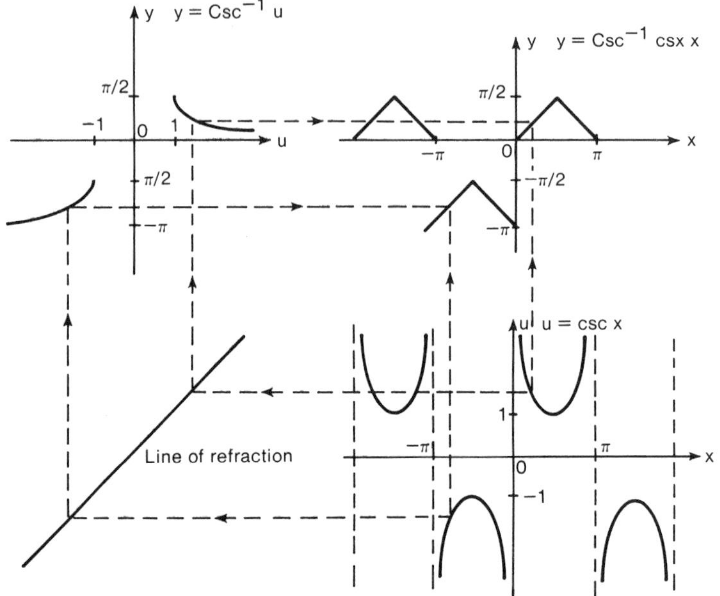

Figure F.6 $y = \text{Csc}^{-1} \csc x$.

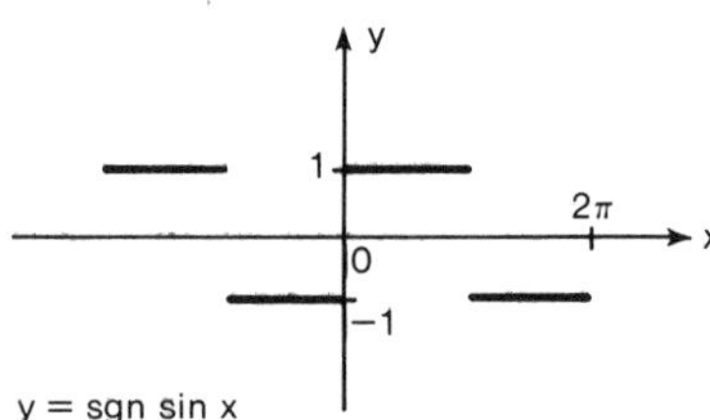

Figure F.7 $y = \text{sgn} \sin x$.

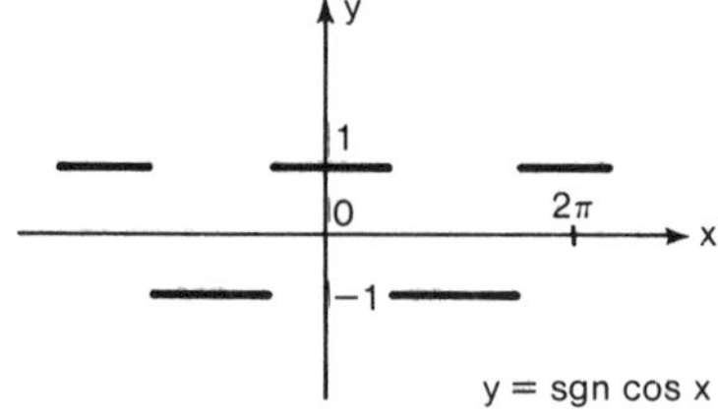

Figure F.8 $y = \text{sgn} \cos x$.

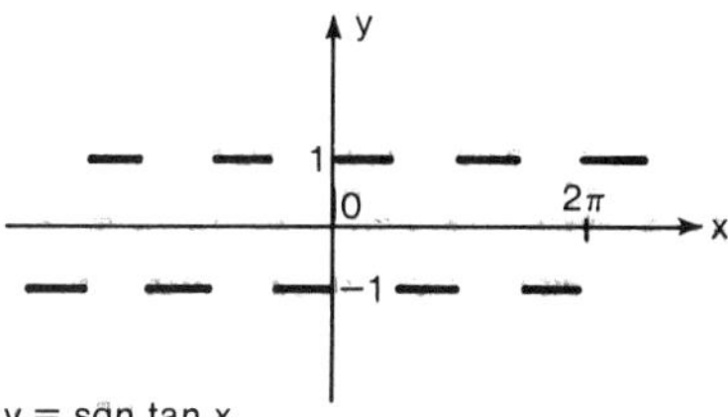

Figure F.9 $y = \text{sgn} \tan x$.

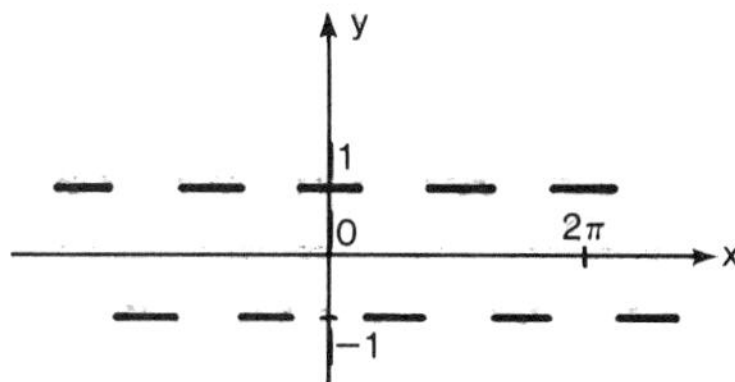

Figure F.10 $y = \text{sgn ctn}\, x$.

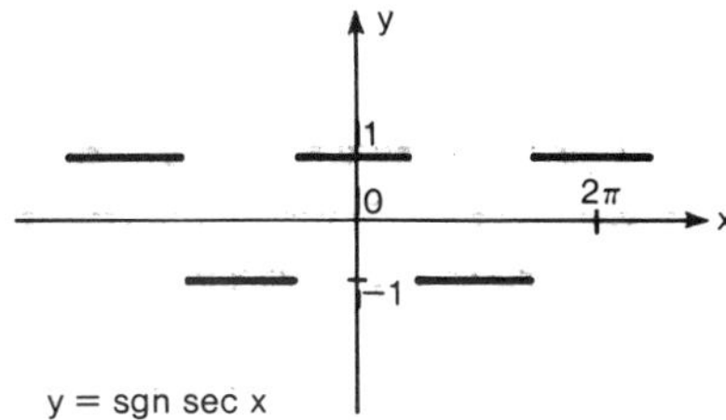

Figure F.11 $y = \text{sgn} \sec x$.

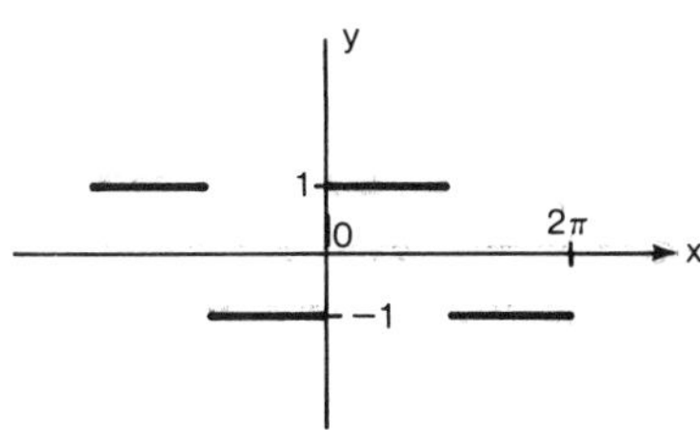

Figure F.12 $y = \text{sgn} \csc x$.

APPENDIX G: FUNCTIONS, OPERATIONS, IDENTITIES

Functions and Operations

A, B, a, b, α, β are constants; u, x are variables.

Sec. Ref.

1. Signum function (sign of a quantity) 2.1

$$\operatorname{sgn} x = x/|x|$$

2. Heaviside step function in terms of the signum function, step at $x = 0$ 2.2

$$S(x) = \tfrac{1}{2} + \tfrac{1}{2}\operatorname{sgn} x$$

3. Sine-derived squarewave 3.2

$$\operatorname{sqr} x = \operatorname{sgn}(\sin x)$$

4. Cosine-derived squarewave 3.2

$$y = \operatorname{sgn}(\cos x)$$

5. Cosine-derived triangular wave 3.3

$$\operatorname{tri} x = \operatorname{Cos}^{-1}\cos x$$

6. Sine-derived triangular wave 3.3

$$y = \operatorname{Sin}^{-1}\sin x$$

7. Tangent-derived linear sawtooth wave 3.3

$$\operatorname{saw}_a x \equiv (a/\pi)\operatorname{Tan}^{-1}\tan(\pi x/a)$$

Sec. Ref.

8. Regular staircase function 3.3

$$\sigma_a(x) \equiv x - \text{saw}_a x$$

10. Pulse function 2.3, 6.1

$$\text{puls}_a^b x \equiv S(x - a) - S(x - b)$$

11. Cosine-derived square pulse train 6.1

$$y = S(\cos x)$$

12. Sine-derived square pulse train 2.5, 6.1

$$y = S(\sin x)$$

13. Triangle-derived square pulse train 6.1

$$y = S(\tfrac{1}{2}\pi - \text{tri } x)$$

14. Rectangular pulse train 6.1

$$y = S(\text{b}\pi - \text{tri } x)$$

15. Clipping operation 7.4

$$\text{clip}_a^b u = \tfrac{1}{2}(a + b) + \tfrac{1}{2}|u - a| - \tfrac{1}{2}|u - b|$$

16. Center clipping operation 7.6

$$\text{ccl}_a^b u = u - \text{clip}_a^b u$$

17. All-wave rectification 7.1

$$\text{rec}_w u = u - u|w| + |u|w$$

w is rectification index

18. Half-wave rectification 7.1

$$\text{rec}_{1/2} u = uS(u)$$

19. Negative half-wave rectification 7.1

$$\text{rec}_{-1/2} u = uS(-u)$$

20. Full-wave rectification 3.1

$$|u| = u \text{ sgn } u$$

21. Pseudorectified sinewave 3.4

$$y = \text{sgn}(\cos x)\sin x$$

22. Pseudorectified cosinewave 3.4

$$y = \text{sqr } x \cos x$$

23. Interrupted cosinewave 3.4

$$y = \cos \frac{\pi(\text{tri } x + \alpha)}{\pi + \alpha + \beta}$$

24. Maximum selector

$$\max(u, v) = uS(u - v) + vS(v - u)$$

Sec. Ref.

25. Minimum selector

$$\min(u, v) = uS(v - u) + vS(u - v)$$

26. Dirac delta function, pulse at $x = 0$ 2.4

$$\delta(x) = dS(x)/dx$$

27. Sine-derived delta pulse train 2.5

$$y = \delta(\sin x)$$

28. Cosine-derived delta pulse train 2.5

$$y = \delta(\cos x)$$

29. Sine-derived alternating delta pulse train 2.5

$$y = \cos x\, \delta(\sin x)$$

30. Cosine-derived alternating delta pulse train

$$y = \sin x\, \delta(\cos x)$$

31. nth power of δ App. B

$$[\delta(x)]^n = |x|^{-n+1}\, \delta(x), \qquad n > 0$$

32. nth order of δ 9.2

$$\delta^n(x) = |x|^{-n+1}\, \delta(x), \qquad -\infty < n < \infty$$

33. Unit amplitude delta function App. B

$$\delta^0(x) = |x|\delta(x)$$

34. Kronecker delta 2.5

$$\delta_{mn} = \delta^0(m - n); \qquad m, n \text{ integers}$$

35. Three-dimensional delta function App. B

$$\delta(x, y, z) = \delta(x)\delta(y)\delta(z)$$

Identities

p is a periodic function.

1. $\operatorname{sgn}(uv) = \operatorname{sgn} u \operatorname{sgn} v$

2. $|u| = (u^2)^{1/2}$

3. $\operatorname{sgn}(\text{saw}_{2\pi} x) = \text{sqr}\, x$

4. $\delta(\text{saw}_{\pi} x) = \delta(\sin x)$

5. $\delta(\operatorname{sgn} x) = 0$

6. $\sin(\text{saw}_{2\pi} x) = \sin x$

7. $\cos(\text{saw}_{2\pi} x) = \cos x$

8. $\tan(\text{saw}_{2\pi}x) = \tan(\text{saw}_{\pi}x) = \tan x$

9. $p(x) = p(\text{saw}_a x)$; $a = \text{period}$

10. $p(x) = p(|\text{saw}_a x|)$; p is even function, $a = \text{period}$

11. $\text{sqr}^2 x = 1$

12. $\text{sgn}^2(\cos x) = 1$

13. $\text{sqr}\, x \sin x = \text{sqr}\, 2x \sin(\text{saw}_{\pi}x)$

14. $\text{sqr}\, x \cos x = \text{sqr}\, 2x|\cos x|$

15. $\text{sgn}(\cos x)(\text{tri}\, x - \frac{1}{2}\pi) = \text{tri}(\text{saw}_{\pi}x) - \frac{1}{2}\pi$

16. $\text{sgn}(\cos x)\text{Sin}^{-1}\sin x = \text{saw}_{\pi}x$

17. $\text{sgn}(\cos x)\text{sqr}\, x = \text{sqr}\, 2x$

18. $\text{sgn}(\cos x)\sin(\text{saw}_{\pi}x) = \sin x$

19. $\text{sgn}(\cos x)\cos(\text{saw}_{\pi}x) = \cos x$

20. $\text{sgn}(\cos x)[\text{tri}(\text{saw}_{\pi}x) - \frac{1}{2}\pi] = \text{tri}\, x - \frac{1}{2}\pi$

21. $\text{sgn}(\cos x)\text{sqr}\, 2x = \text{sqr}\, x$

22. $\text{saw}_{2\pi}x - \frac{1}{2}\,\text{sqr}\, x = \text{saw}_{\pi}x - \frac{1}{2}\pi\,\text{sqr}\, 2x$

23. $\text{sqr}\, 2x + \dfrac{2}{\pi}\sigma_{\pi}(x) = \text{sqr}\, x + \dfrac{2}{\pi}\sigma_{2\pi}(x)$

24. $\text{Cos}^{-1}\sin x = \text{tri}(x + \frac{1}{2}\pi)$

25. $|\cos x| = \cos(\text{saw}_{\pi}x)$

26. $\sigma_{\pi}(x - \frac{1}{2}\pi) = \dfrac{\pi}{2}\,\text{sqr}\, 2x + \sigma_{\pi}(x) - \dfrac{\pi}{2}$

27. $|\text{saw}_{2\pi}x| = \text{tri}\, x$

28. $\text{Sin}^{-1}\sin(\text{saw}_{\pi}x) = \text{saw}_{\pi}x$

29. $\text{Sin}^{-1}\sin x = \text{tri}(x + \frac{1}{2}\pi) - \frac{1}{2}\pi$

30. $\text{Ctn}^{-1}\text{ctn}\, x = \text{saw}_{\pi}(x + \frac{1}{2}\pi) + \frac{1}{2}\pi$

31. $\text{Csc}^{-1}\csc x = \text{Sec}^{-1}\sec(x + \frac{1}{2}\pi) + \frac{1}{2}\pi$

32. $\text{sgn}(\cos x) = \text{sqr}(x + \frac{1}{2}\pi)$

33. $\text{sgn}(\tan x) = \text{sqr}\, 2x$

34. $\text{sgn}(\sec x) = \text{sqr}(x + \frac{1}{2}\pi)$

35. $\operatorname{sgn}(\csc x) = \operatorname{sqr} x$

36. $\delta(2 \sin \tfrac{1}{2}x) - \delta(2 \cos \tfrac{1}{2}x) = \delta(\sin x)\cos x$

37. $\operatorname{saw}_a x = \dfrac{a}{2\pi} \operatorname{saw}_{2\pi}(2\pi x/a)$

38. $\operatorname{sgn}(\cos x)\operatorname{saw}_\pi x = \operatorname{Sin}^{-1}\sin x$

39. $\operatorname{sgn}(\cos x)\sin x = \sin(\operatorname{saw}_\pi x)$

Binary Digital Logic Operations in Terms of the Functions of This Volume

a and b take on only the values 0 and 1.

δ_{ab} is the Kronecker delta.

1. NOT-gate (inverter)

$\delta^0(a) = 1 - a$

2. AND-gate

ab

3. NAND-gate

a, b: $\delta^0(ab) = 1 - ab$

4. OR-gate

a, b: $\operatorname{clip}^1(a + b) = a + b - ab$

5. NOR-gate

a, b: $\delta^0(a + b)$

6. XOR-gate (exclusive-or gate)

a, b: $|a - b| = a + b - 2ab$

7. XNOR-gate (exclusive-nor gate)

a, b: $\delta^0(a - b) = \delta_{ab}$

APPENDIX H: DERIVATIVES

These are complete derivatives; *a* and *b* are constants; *u* and *x* are variables.

	Sec. Ref.
1. $\dfrac{d}{dx}S(u) = \delta(u)\dfrac{du}{dx}$	4.2
2. $\dfrac{d}{dx}\text{sgn}\ u = 2\delta(u)\dfrac{du}{dx}$	4.2
3. $\dfrac{d}{dx}\lvert u\rvert = \text{sgn}\ u\dfrac{du}{dx}$	4.1
4. $\dfrac{d}{dx}\text{tri}\ u = \text{sqr}\ u\dfrac{du}{dx}$	4.3
5. $\dfrac{d}{dx}\lvert\cos u\rvert = -\text{sgn}(\cos u)\sin u\ \dfrac{du}{dx}$	4.1
6. $\dfrac{d}{dx}\lvert\sin u\rvert = \text{sqr}\ u\ \cos u\ \dfrac{du}{dx}$	4.1
7. $\dfrac{d}{dx}\text{sqr}\ u = 2\cos u\ \delta(\sin u)\ \dfrac{du}{dx}$	4.3
8. $\dfrac{d}{dx}\ \text{saw}_a u = 1 - a\delta\left(\dfrac{a}{\pi}\cos\dfrac{\pi u}{a}\right)\dfrac{du}{dx}$	4.3
9. $\dfrac{d}{dx}\text{saw}_a^2 u = 2\ \text{saw}_a u\ \dfrac{du}{dx}$	

Sec. Ref.

10. $\frac{d}{dx}\text{tri}^2 u = 2 \text{ tri } u \text{ sqr } u \frac{du}{dx} = 2 \text{ saw}_{2\pi} u \frac{du}{dx}$

11. $\frac{d}{dx}\sigma_a(u) = a\delta\left(\frac{a}{\pi}\cos\frac{\pi u}{a}\right)\frac{du}{dx}$ 4.3

12. $\frac{d}{dx}\text{clip}_a^b u = \text{puls}_a^b u \frac{du}{dx}$ 7.4

13. $\frac{d}{dx}[f(u)\text{puls}_a^b u]$

$$= \left\{\text{puls}_a^b u \frac{d}{dx} f(u) + f(u)[\delta(u - a) - \delta(u - b)]\right\}\frac{du}{dx}$$

APPENDIX I: INTEGRALS

A, B, a, and b are constants; u and x are variables. An arbitrary constant should be added to each integral.

	Sec. Ref.
1. $\int \delta(x)\, dx = S(x)$	2.4
2. $\int f(x)\, \delta(x - a)\, dx = f(a)S(x - a)$; $f(x)$ single valued at $x = a$	2.5
3. $\int \delta(\cos x)\, dx = (1/\pi)\sigma_\pi(x)$	2.5
4. $\int \operatorname{sgn} x\, dx = \lvert x\rvert$	5.2
5. $\int \lvert x\rvert\, dx = \dfrac{1}{2}\lvert x\rvert x$	5.2
6. $\int f(x) \operatorname{step}_a x\, dx = \operatorname{step}_a x \int_a^x f(x)\, dx$; $f(x)$ continuous at $x = a$	5.2
7. $\int \operatorname{saw}_a x\, dx = \dfrac{1}{2} \operatorname{saw}_a^2 x$	5.3
8. $\int \operatorname{sqr} x\, dx = \operatorname{tri} x$	5.3
9. $\int \lvert\cos x\rvert\, dx = \operatorname{sgn}(\cos x)\sin x + (2/\pi)\sigma_\pi(x)$	5.3
10. $\int \lvert\sin x\rvert\, dx = -\operatorname{sqr} x \cos x + (2/\pi)\sigma_\pi(x - \tfrac{1}{2}\pi)$	5.3
11. $\int \operatorname{saw}_a^2 x\, dx = \dfrac{1}{3} \operatorname{saw}_a^3 x + \dfrac{1}{12} a^2 \sigma_a(x)$	5.5
12. $\int x \operatorname{saw}_a x\, dx = \dfrac{1}{2} x \operatorname{saw}_a^2 x - \dfrac{1}{6} \operatorname{saw}_a^3 x - \dfrac{1}{24} a^2 \sigma_a(x)$	5.5

Sec. Ref.

13. $\int \sigma_a^2(x)\,dx = \frac{1}{3}x^3 - x\,\text{saw}_a^2 x + \frac{2}{3}\text{saw}_a^3 x + \frac{1}{6}a^2\sigma_a(x)$ 5.5

14. $\int \text{puls}_a^b x\,dx = \frac{1}{2}|x - a| - \frac{1}{2}|x - b|$ 6.1

15. $\int f(x)\text{puls}_a^b x\,dx = \text{puls}_a^b x \int f(x)\,dx + [S(x-u)\int f(u)du\,]_{u=a}^{u=b}$ 6.2

16. $\int \text{tri}\ x\,dx = \frac{1}{2}\text{tri}\ x\ \text{saw}_{2\pi}x + \frac{1}{2}\pi\sigma_{2\pi}(x)$ 5.3

17. $\int \delta^n(x)\text{sgn}\ x\,dx = \frac{-1}{n-1}\,\delta^{n-1}(x), \qquad n \neq 1$

18. $\int \delta(x)\text{sgn}\ x\,dx = \text{Ln}|x|\ \delta^0(x)$

19. $\int |u|\,dx = x|u| - \text{sgn}\ u \int_0^u x\,du$

APPENDIX J: TERMS

Electronic Term	*Mathematical Term*
a-c component	a periodic function less its average value
a-c log	inverse hyperbolic sine
amplification (gain $= \alpha$)	multiplication by a constant α
amplitude	maximum value
amplitude modulation	multiplication by a variable
attenuation (loss $= \beta$)	division by a constant β
compression	an operation that changes a linear function to one whose graph is concave downwards
d-c component	the average value of a periodic function
drift	the addition of a slowly varying quantity
effective value	(same as rms)
expansion	an operation that changes a linear function to one whose graph is concave upwards
flyback	steeper sloping portion of a sweep function
full-wave rectification	absolute-value operation
function generation	functional operation
input	independent variable
inversion, analog	finding the negative
inversion, digital	finding the 1's complement

Electronic Term	*Mathematical Term*
offset	addition of a constant
output	dependent variable
parallel operation, ‖	$a \parallel b = ab/(a + b)$
phase inversion	(same as inversion, analog)
retrace	(same as flyback)
rms (root mean square)	rms; sometimes rms of a sinewave
sweep function	asymmetrical triangular wave (usually)
switching function	step function
trace	shallower-sloping portion of a sweep function
transfer characteristic	operation
true rms	rms
zero-crossing detection	signum operation on a continuous function

APPENDIX K: SYMBOLS

Symbol	Meaning	Sec. Where Introd.
1. $\lim_{x \to a+}$, $\lim_{x \to a-}$	unilateral limits	1.3
2. $F(a + 0)$, $F(a - 0)$	unilateral limits	4.2
3. $\underset{\leftarrow}{\leq}$, $\underset{\rightarrow}{\leq}$	unilateral inequalities	3.3
4. $(\ )^{1/n}$	principal value of nth root	1.2
5. $\rightarrow\rightarrow\rightarrow$	flag	1.6
6. $((\ ))$	flag	5.4
7. Δ	a general delta function	5.4
8. Δx, Δy, ΔS	increment of x, y, or S	
9. $(\ )'$ (prime)	**a.** a tag	2.5
	b. 1's complement	7.6.2
	c. coefficient of imaginary part of complex quantity	11.1
	d. formal derivative of δ	App. B
10. $(\dot{\ })$ (dot)	time derivative	4.4
11. C	a constant	2.4

APPENDIX L: BIBLIOGRAPHY

Nonlinear Circuits Bibliography

1. Analog Devices Engineering Staff, D. H. Sheingold, ed., *Nonlinear Circuits Handbook* (2nd ed.). Norwood, Mass.: Analog Devices, Inc., 1976. Order from Analog Devices, Inc., P.O. Box 796, Norwood, MA 02062. Library of Congress Catalog Card No. 75-42559. This publication picks up where Philbrick (see #14) leaves off. Contains an extensive bibliography of articles in trade journals.
2. Chance, B., V. Hughes, E. F. MacNichol, D. Sayre and F. C. Williams, *Waveforms*. New York: McGraw-Hill, 1949. A pioneering, comprehensive work on nonlinear circuits.
3. Fullagar, D., *Understanding and Applying the Analog Switch*. Cupertino, Cal.: Intersil, Inc., 1978.
4. Giles, J. N., *Fairchild Semiconductor Linear Integrated Circuits Applications Handbook*. Mountain View, Cal.: Fairchild Semiconductor, 1967. Library of Congress Catalog Card No. 67-27446.
5. Hufault, J. R., *Op Amp Network Design Manual*. Tucson, Ariz.: Marketpul Corp., 1982. Order from Marketpul Corp., 500 E. San Moritz, Tucson, AZ 85704. Library of Congress Catalog Card No. 82-60678. Contains many circuits with component values.

6. Korn, G. A., and T. M. Korn, *Electronic Analog and Hybrid Computers* (2nd ed.). New York: McGraw-Hill, 1972.

7. ———, *Electronic Analog Computers* (2nd ed.). New York: McGraw-Hill, 1956. Library of Congress Catalog Card No. 56-8176. A classic reference.

8. Landee, R. W., D. C. Davis and A. P. Albrecht, *Electronic Designers' Handbook*. New York: McGraw-Hill, 1957. Library of Congress Catalog Card No. 56-6898. A classic work containing several chapters on nonsinusoidal oscillators and clippers.

9. Lenk, J. D., *Handbook of Simplified Solid State Circuit Design*. Englewood Cliffs, N.J.: Prentice-Hall, Inc., 1971. Library of Congress Catalog Card No. 75-149976.

10. Meiksin, Z. H., and P. C. Thackray, *Electronic Design with Off-the-Shelf Integrated Circuits* (2nd ed.). Englewood Cliffs, N.J.: Prentice-Hall, Inc., 1984.

11. Oxner, E. S., *Power FETs and Their Applications*. Englewood Cliffs, N.J.: Prentice-Hall, Inc., 1982. Many switching applications are described.

12. Paynter, H. M., ed., *A Palimpsest on the Electronic Analog Art*. Boston, Mass.: Geo. A. Philbrick Researches, Inc., 1955. A classic, now out of print.

13. Phelps, R. S., ed., *750 Practical Electronic Circuits*. Blue Ridge Summit, Penn.: Tab Books, Inc., 1983. Library of Congress Catalog Card No. 82-5988.

14. Philbrick Researchers Engineering Staff, *Applications Manual for Computing Amplifiers* (2nd ed.). Boston, Mass.: Nimrod Press, Inc., 1966. Library of Congress Catalog Card No. 66-19610. A pioneering work on operational amplifiers in linear and nonlinear circuits. Reprinted in 1984 as *Applications Manual for Operational Amplifiers,* Teledyne Philbrick Microcircuits, Allied Drive at Route 128, Dedham, Mass. 02026.

15. Pippenger, D. E., and C. L. McCollum, *Linear and Interface Circuit Applications*. Dallas, Tex.: Texas Instruments, Inc., 1974.

16. Reich, H. J., *Functional Circuits and Oscillators*. Cambridge, Mass.: Boston Technical Publishers, Inc., 1965.

17. Savant, C. J., Jr., *Basic Feedback Control System Design*. New York: McGraw-Hill, 1958. Library of Congress Catalog Card No. 57-10228. Also contains information on the Laplace transform.

18. Wait, J. V., L. P. Huelsman and G. A. Korn, *Introduction to Operational Amplifier Theory and Applications*. New York: McGraw-Hill, 1975.

19. Wong, Y. J., and W. E. Ott, *Function Circuits, Design and Applications,* The Burr-Brown Electronics Series. New York: McGraw-Hill, 1976.

Mathematics Bibliography

Source material for the Heaviside step function, the Dirac delta function, and the Laplace transform.

1. Arfken, G., *Mathematical Methods for Physicists* (2nd ed.). New York: Academic Press, 1973. Library of Congress Catalog Card No. 73-119611.
2. Churchill, R. V., *Modern Operational Mathematics in Engineering*. New York: McGraw-Hill, 1944. This book is the definitive source for the Laplace transform.
3. Korn, G. A., and T. M. Korn, *Mathematical Handbook for Scientists and Engineers*. New York: McGraw-Hill, 1961. Library of Congress Catalog No. 59-14456.
4. Mathews, J., and R. L. Walker, *Mathematical Methods of Physics* (2nd ed.). New York: W. A. Benjamin, Inc., 1970. Library of Congress Catalog Card No. 71-80659.

INDEX

T